SPACE TOURISM

SPACE TOURISM

Adventures in Earth Orbit and Beyond

Michel van Pelt

Copernicus Books   *In Association with*   Praxis Publishing Ltd.

An Imprint of Springer Science+Business Media

Published in the United States by Copernicus Books,
an imprint of Springer Science+Business Media.

Copernicus Books
37 East 7th Street
New York, NY 10003
www.copernicusbooks.com

Library of Congress Cataloging-in-Publication Data

Van Pelt, Michel.
    Space tourism: adventures in Earth orbit and beyond / Michel van Pelt.
       p. cm.
    Includes bibliographic references and index.
    ISBN 0-387-40213-6 (alk. paper)
1. Space tourism—Popular works. 2. Astronautics—Popular works. I. Title.

    Tl793.V293 2005
    910'.919—dc22   2004059007

Manufactured in the United States of America.
Printed on acid-free paper.

9 8 7 6 5 4 3 2 1

ISBN 0-387-40213-6   SPIN 10932766

To Ria and Leo

CONTENTS

Contents

Contents

PREFACE

Many scientific papers and popular articles have been written on the topic of space tourism, describing anything from expected market sizes to the rules of three-dimensional microgravity football. But what would it actually feel like to be a tourist in space, to be hurled into orbit on top of a controlled explosion, to float around inside a spacecraft and be able to look down on your home town from above the atmosphere?

Most of us love the view from high places. The Euromast tower in The Netherlands features a "space ride" elevator to its top. Following a short countdown, white smoke billows out from the bottom of the elevator and the thunder of a Space Shuttle launch booms in your ears. The rather slow climb up is somewhat of an anticlimax, but the tower does offer a spectacular view of the city of Rotterdam. What if the mast did not end at 185 meters? What if the elevator continued up for another 200 kilometers and really went into space? What if you could see half the planet instead of just part of a city, no matter what the weather is like?

The book you are holding now is for those people who wonder about this, who dream about circling the Earth every 90 minutes, freely floating in front of a large window, looking down at the blue planet beneath. It is also for those who have never dreamed about this but are curious to know why others do.

This book takes you on a future space tourism trip, and along the way explains such things as the required spacecraft technology, medical issues, astronaut training and the possibilities for holidays far from Earth.

The Introduction explains what space tourism is and why the time has arrived to take it seriously. Starting with "Before the Flight", the next three chapters explain the medical and safety issues and show the current developments in space and space-related tourism.

"Launch" and "The Sky is not the Limit" investigate the technology and costs of current and future launch vehicles, and how space-tourism vehicles could be made cheap enough to enable a mass space-tourism market. "In Orbit" and "Space Stations" deal with life in microgravity, space stations and the fun things that a trip in Earth orbit offers to space tourists. "The End of the Tour" and "Returning from Space" are about returning to Earth and the effects of a return to gravity on the human body that has become used to living in space. "To the Moon, Mars and Further" looks into the more distant future, when space tourists may leave Earth orbit to explore the rest of the solar system and beyond.

As you will see, space tourism is no longer science fiction, but merely a logical next step in the evolution of spaceflight. Accidents can still happen, as the tragic loss of the crew of the Space Shuttle Columbia recently proved. No technology can ever be made 100 percent safe. Nevertheless, we have learned how to get a few people into orbit and back, and how to keep them safe and comfortable in the cold, dark hostility of space. We even managed to land some of them on the Moon. Now we are about to open up space for more and more people, creating a routine and safe access to it that will enable thousands, tens of thousands and eventually even millions to travel beyond Earth's atmosphere. The drive behind this is one of the world's largest, most powerful economic forces: tourism.

I want to thank Clive Horwood and Dr John Mason of Praxis, Alex Whyte (Praxis copy editor), and Paul Farrell and Anna Painter of Copernicus, for their support to this book and their helpful comments and suggestions. For moral support during the writing of this book, I thank my girlfriend Stefania Monni. Especially helpful were my friends from the International Space University's SSP 2000 Space Tourism project, whose research, imagination and enthusiasm form the basis for this work. To Sven Abitzsch, Alessandro Atzei, Torsten Bieler, Peter Buist, Dennis Gerrits, Hervé Joumier, Zeina Mounzer, Ron Noteborn, Paulo Ponzio, Rogier Schonenborg, Benjamin Schreiber, Stella Tkatchova and Paloma Villar I express my thanks for the information and inspiring conversations about space and many other things, which brought out countless ideas. Also thanks to my fellow editors of the Dutch space magazine *Ruimtevaart* and editor Gerard Keijzers of *Astruim*: preparing articles for these publications has taught me how to put pen to paper. Finally, thanks to Marieke van Hijum for supplying the following remark by Wilbur Wright, said shortly after the Wright brothers' first flight in 1903:

"It is not really necessary to look too far in the future;
we see enough already to know that it will be magnificent.
Only let us hurry and open the road."

INTRODUCTION

PRIL 19, 2004. Three years after 2001, the year for which Arthur C. Clarke and Stanley Kubrick predicted comfortable routine flights to Earth orbit, space hotels and moonbases in their movie *2001: A Space Odyssey*, a Soyuz rocket lifts off from the Baikonur launch site in Kazakhstan (see Figure 1).

People at launch control and at various space agencies around the world are anxiously watching the rocket's flight, as they have done for decades. In the evening, the successful launch will be on the news, as the news anchor readers will tell the public of how yet another crew of bold spacefarers has left the planet. But the International Space Station the astronauts are going to dock to is nothing like the giant wheel depicted in the movie. It is a relatively small cluster of very expensive modules that can house no more than three crew members. In concept, the latest space outpost is not much different from the first orbital stations launched three decades earlier, while the venerable Soyuz rocket is based on the launcher that put the first man into space in 1961.

Dutch astronaut André Kuipers, working for the European Space Agency, is strapped in a capsule of the type that first flew almost 40 years ago. His selection and that of his two fellow crew members was rigorous; they had to pass many demanding physical and psychological tests to get one of the most privileged jobs in the world. The subsequent training took years, with all the procedures and technology that had to be mastered. However, for them the experience of the flight, the responsibility of the mission, the view of Earth from 250 kilometers (155 miles) up and the experience of microgravity are all worth it.

FIGURE 1   *A Soyuz launcher thunders into the heavens to deliver a crew of three cosmonauts into orbit. It already flew two space tourists to the International Space Station on two separate missions. [Photo: ESA]*

ORBITING THE EARTH

The launch of a crewed space rocket is still a breathtaking piece of human intellect and technical capability. It is also still rather expensive. A ride on top of a Space Shuttle or Soyuz is normally only available to a few highly trained, carefully selected champions.

Isn't it strange that after more than four decades, human spaceflight is still regarded as something very special? Space remains something only a small, elite group of astronauts is allowed to experience. Until now, only some 440 Western astronauts, Russian cosmonauts and Chinese yuhangyuans have flown in space, and only 12 of them have set foot on the Moon. In contrast, tickets were sold for fun rides in airplanes almost immediately after the Wright brothers got their Flyer, the world's first motorized airplane, off the ground.

Unlike air travel, getting into Earth orbit is currently a risky and very expensive undertaking. Astronauts therefore only get launched for serious reasons, such as conducting scientific experiments or building space stations. They have to invest in years of training to get ready for a flight that may last only a week or so, and once they are in orbit there is little time in the busy flight schedule to admire the views or have some fun with microgravity.

Many "ordinary" people are fascinated by space but had no hope of ever going there themselves ... until now. After nearly half a century of spaceflight, things finally seem about to change, and the first tourists are taking short vacations on board the International Space Station. The idea of space tourism is starting to be taken serious by aerospace companies, space agencies and the tourism industry.

Space has much to offer the adventurous traveler: a thrilling ride into orbit on the thrust of roaring rocket engines, a magnificent view of Earth, and the unique feeling of freedom while floating in weightlessness. For the near future you can add the glory of being a pioneer to the attraction – the excitement of going where no tourist has gone before. In the years to come, the fun could even include such spectaculars as dune buggy riding on the Moon or mountaineering on Mars.

Astronauts say it is impossible to describe the grandeur of the Earth as they have seen it (see Figure 2). From orbit, you see whole mountain ranges, volcanoes, entire glaciers, storms, ocean currents – endless amounts of detail on a very intriguing planet. Seeing the entire Earth, otherwise impossible to achieve in a lifetime, can now be done in a few days. Circling around our beautiful planet you can enjoy 16 sunrises and 16 sunsets in the same day, and they are all perfect.

FIGURE 2 *A crew member on board the Space Shuttle photographs the Earth through a window. The view of our planet from orbit is one of the main attractions of a spaceflight. [Photo: ESA]*

The presence of humanity, so obvious in all the heavily populated areas on the ground, seems hardly noticeable. Our cities and large structures, like dams and bridges, appear to be nothing compared to the vast polar caps or the magnificence of the Himalayas. However, on the night side of the planet lights from villages, towns and cities clearly reveal how extensively people have taken over the planet.

From the Moon, Apollo astronauts looked at that small, blue marble that they could hold between their fingers, and marveled about all the history, technology, wars, human successes and failures and everything else that had occurred there. On the ground we are lured into the perception that the Earth is endless, eternal and the biggest thing in the universe. From space, however, our planet reveals itself to be a tiny oasis in the vast hostility of space.

From space you cannot see geographical borders and you begin to wonder about all the wars and misery those artificial lines have created. You start to wonder how people can destroy vast parts of the Amazon forest and ignore its effect on our world, while it is so clear that the Earth is just a small, fragile dot of life in a very large, very empty and very indifferent universe. What seems divided and vast in an atlas now appears united on one rather limited sphere.

Because of human spaceflight, for the first time people can be truly detached from the Earth and look at the planet from a completely new point of view. Many astronauts say that their flight had a deep impact on their thinking, that they felt a stronger bond with the planet and the whole of humanity ever since. NASA astronaut Roberta Bondar said about this: "To fly in space is to see the reality of Earth, alone. The experience changed my life and my attitude towards life itself. I am one of the lucky ones." And according to NASA astronaut Joseph Allen: "With all the arguments, pro and con, for going to the Moon, no one suggested that we should do it to look at the Earth. But that may in fact be the most important reason."

Professional astronauts, mostly pilots, doctors and engineers, expressed the hope that one day poets and writers could go up in orbit to try to describe the marvelous sight. Many of them also think that if more people, and especially politicians, could see our planet from space, we would have fewer wars and care more for our environment. Space tourism will make it possible, as it offers this unique experience to more and more people of all walks of life.

Preliminary market surveys indicate that the number of people willing to spend serious money on an opportunity to orbit the Earth is huge. This is not surprising, as adventurous holiday activities such as wild water rafting, scuba safaris, trekking in the Himalayas and cruises to the Arctic are becoming ever more popular.

And space is popular too. The NASA Mars Pathfinder website recorded 556 million hits when the tiny Sojourner rover took its little stroll over the surface of Mars. In 1998, an estimated 2 million people traveled to Cape Canaveral to see America's first orbital astronaut John Glenn embark on his second space flight. More recently, NASA's Mars Excursion Rovers made front page news several times and their pictures of the red planet appeared in thousands of magazines, television shows, websites and books.

For people who would like to get a more direct feel for what it is like to be an astronaut, such companies as Space Adventures and Incredible Adventures offer astronaut training camps, rides in Russian military jets to the edge of the atmosphere and spacewalk experiences in large swimming pools. Other companies are planning to launch privately funded space missions such as a moonrover that can be operated via the Internet.

The X-Prize organization is offering $10 million to the first team that is able to launch a 3-person rocket vehicle to an altitude of 100 kilometers (62 miles), and then do this again within two weeks using the same vehicle. The X-Prize competition is expected to result in reusable rocket planes able to take paying customers on a suborbital trajectory into space, where they can enjoy the view and microgravity for a few minutes. The technology for such flights has existed since the early 1960s, when the X-15 fully reusable rocket planes reached altitudes over 100 kilometers (62 miles) and could be relaunched within a week.

But the ultimate experience is, of course, a stay in Earth orbit for at least a day, giving ample time to view your home planet as you circle around it once every 90 minutes.

As early as 1988, the famous singer John Denver was interested in buying a flight to the Soviet Mir space station. In the end he declined because of the $10 million ticket price and the long training period required.

In 1990 Japanese journalist Toyohiro Akiyama was sent to Mir by his employer, the television station TBS, for $12 million. Although he cannot be regarded as the first space tourist, because he did not go just for recreation and did not pay for the trip himself, his flight demonstrated that ordinary, relatively untrained people can go into space without too much trouble.

The first British citizen in space was Helen Sharman, who was launched into orbit as part of a Russian Soyuz crew. Sharman was a research technologist for a confectionery company when she was selected for cosmonaut training by an organization called Antequera Ltd. This company was based in London but was fully owned by a Moscow bank. It organized the flight to strengthen the ties between the Soviet Union and the UK, but hoped to raise sufficient money from sponsors to cover the cost. However, interest in the flight from private sponsors and the British government was very low and even while Sharman was still in training, Antequera was formally dissolved. The Soviet bank that owned the company decided, however, to sponsor the mission itself in the interest of generating pro-Soviet publicity, and in 1991 Sharman was sent into orbit. Sharman's flight was not a fully private enterprise, but nor was it a normal government mission.

The real first space tourist was launched on April 28, 2001, on board a Russian Soyuz capsule on top of a rocket with the same name. Dennis Tito's flight took him to the International Space Station (ISS), where he and his crew spent a week before returning to Earth. They came back in the older Soyuz that was already docked to the station, leaving a fresh one as the new lifeboat for the ISS. Self-made millionaire Tito paid some $20 million for his flight. It was only a small part of his estimated wealth of $200 million, but was still a lot of money to pay for a holiday.

The Russians could make good use of the money to finance their commitment to the Space Station, as their entire budget for the year was scarcely seven times higher than the price Tito paid. However, ISS partners NASA, the European Space Agency (ESA), the Japanese Space Agency and the Canadian Space Agency were less thrilled with having a tourist on board their costly outpost, but they could not prevent the Russians from flying Tito to the station, as each partner is allowed to select its own crews. NASA could only insist on Tito agreeing not to sue the space agency or its partners in the event of personal injury. He would also have had to pay for any damage he caused during the flight.

Despite all of this, Tito enjoyed his flight to the maximum, listening to opera music, shooting video and stereographic pictures of the Earth, and floating from one part of the station to another. Tito was the first astronaut ever to go into space without a full list of tasks; he had eight days to just experience space to the fullest. His first words when he entered the Space Station were "I love space."

A year later 28-year-old South African Internet millionaire Marc Shuttleworth followed in Tito's footsteps. By that time, an agreement had been reached between the Russians and the other ISS partners concerning space tourist trips to the Station. There was no open resistance to the flight.

Mark's official designation became "Flight Participant" – a name more fitting to the serious world of government space programs than the frivolous "space tourist." Mark liked the new name much better too; instead of being merely a passenger, he would be conducting a number of experiments for South African institutes and universities and also helping his fellow crew members with some tasks. He would be more than a tourist.

Ten days after blasting off the Baikonur launch pad, and after a thrilling period of Earth-gazing, weightlessness and experiments, Shuttleworth returned to Earth with a whole new view of the world. It was a view he wanted to share with the children of South Africa so, together with a rapper, a DJ and a graffiti artist, he toured around the country, visiting some 50 schools. By telling them about his experiences in orbit, he strove to excite the children and make them enthusiastic about space and a career in science.

Although extensive space tourism market studies are sorely needed to enable reliable forecasts and to attract investors, it is expected that the space tourism business could grow to be economically very important. Indeed, it could be much more important than any other space application such as telecommunications satellites or remote-sensing applications. After all, current terrestrial adventure travel is a $120 billion per year industry – many times the budgets of all the space agencies in the world put together. And tourism as a whole is a much larger business, with an annual world market volume of about $3,400 billion!

An orbital space tourism business could also provide a boost to the launcher industry, which is now solely dependent on the relatively small and only slowly expanding satellite launch market.

America, China, Europe, Japan and Russia are still leaders in the development of heavy launchers, an advantage they can use to tap into the space tourism market. They could develop reusable spaceplanes with the low cost per flight that this requires. At the moment we are stuck with expensive conventional rockets, very similar to those that were used by America and the Soviets to win the Cold War space race. Large rockets, developed to project nuclear warheads, were available to both super powers at the time. Using these to launch satellites and crewed capsules made rapid achievements possible. The military missiles, however, were never developed for the economical use of space.

High launch costs limit the benefits we can obtain from space applications. Until now only the telecom business and some remote-sensing satellite operators have been able to make serious money from the use of space. Because of this, only a limited number of satellites need to be launched each year, for which the expensive and expendable launchers currently in use are quite sufficient. The incentive to develop reusable launchers capable of launching many more satellites for much lower prices is therefore rather low.

Space tourism could help us to break out of this cycle. Low-cost launchers developed for the space tourism industry would enable other new space applications too, such as solar-powered satellites for collecting solar energy and beaming it to Earth, or orbital factories for new materials that can only be produced in space.

Spaceplanes that can put things into orbit can also be used for fast, suborbital intercontinental flights. With such vehicles, no place on Earth would be more than an hour's flight away.

Cheap access to space could also make futuristic projects like lunar mining and the colonization of the solar system much more feasible. In this way, space tourism would not just be a nice spin-off from the "more serious" use of space but would actually be an enabler for all sorts of new opportunities.

Moreover, space tourism can boost public support for space exploration in general. Part of the popularity of the 1960s Apollo Moon program was that people thought it would soon be possible to make trips to the Moon themselves. Spaceflight may now finally fulfill this promise, and benefit from increased public and political support and funding in the process.

In the short term, new technologies and industries like those related to space tourism may give an economic boost to the participating countries.

In the longer term, many space visionaries predict that the human race will stagnate or even collapse if we remain an earthbound society for ever. Without the new technologies, the increased scientific understanding, and the inspiration resulting from space exploration we may become disinterested, demoralized and rooted.

We could find ourselves trapped on an increasingly polluted planet where resources such as metals and fuels become increasingly scarce. If the population growth continues at its current rate, we may at some point have insufficient space and resources on this planet to guarantee a reasonable life for everyone. In fact, some argue that we have already reached this situation as large numbers of people around the world live in poverty.

Space exploration will give us access to the riches of the solar system: asteroids full of metals, endless amounts of room for industrial expansion and planets and moons to colonize. Pollution-producing factories could be sent into space, leaving the Earth as an ideal, green and healthy place to live.

Others point out that the impact of a large comet or asteroid could mean the end of us if we remain a one-planet civilization. In the past such impacts resulted in large amounts of dust in the atmosphere, blocking sunlight for years and killing off the dinosaurs. Recent investigations suggest that large impacts have been more common in our past than we have realized until now, sometimes pushing close to 90 percent of all living beings on Earth into extinction. If we want to ensure the survival of humanity, it may be necessary for some of us to live on the Moon, Mars or even in artificial space colonies.

From a more philosophical point of view, some argue that life, and especially humanity, has always pushed its boundaries. Fish crawled on land, walking reptiles became birds and took to the sky, and apes living in trees started to walk upright and eventually flew into space. Exploration is in our blood and is something we yearn to do; it defines us as human beings. Without the possibility for groups of people to break away from global society and try new ways of living, and without the "frontier" spirit, humanity may find itself at an evolutionary dead-end.

Space tourism may be the enabler of our next step in the evolution from an Earth-based to a solar-system-based civilization. It may result in much lower costs for access to space and human spaceflight becoming part of everyday life. Space tourism could be essential for us to become an interplanetary species, and may thereby be crucial for our survival.

This book explores the possibilities of space tourism. It is about taking a fun ride into orbit, about what it could be like to take a cruise in space 15 or 20 years from now. Who could go? What kind of training would be required? What kind of experiences may future space tourists (maybe you!) enjoy?

Some sections of this book take a leap into the future, and follow you as a future space tourist, going through astronaut selection, training, the launch, the flight in space and your return to Earth. These parts are intended to show you the expected possibilities from the point of view of a customer of a future spaceflight tour operator.

As a comparison, the past and current human spaceflight situation is described, to show how ongoing developments can make the space tourism dream a reality.

Fasten your seatbelts, start the countdown, and get ready for launch. . . .

1

BEFORE THE FLIGHT

THE experience of a future space tourist's trip into orbit will begin with tests, because embarking on an orbital flight will be very different from anything you have previously experienced on Earth. There are the relatively high acceleration forces of the launch, the response of your body to the microgravity conditions and the subsequent reactions to the return to Earth, all of which you need to prepare for. Then you must learn the peculiarities of life in orbit. Furthermore, space is a dangerous environment and there is little room for mistakes, so critical procedures need to be practiced.

There will be medical checks to ensure that the flight will not be a hazard for your health. And training will be required to make you familiar with all the procedures, onboard equipment and safety regulations and get you ready for life in space. However, these procedures will be very different from those for astronauts today....

THE EXPERIENCE BEGINS

As you walk into the Space Medicine wing of the hospital, the slight hint of antiseptic that always seems to hang around in hospital corridors reminds you of some past experiences in this building. You really do not like that smell! But today you are not here for an appendix operation or a broken leg, but to take the first step of a trip into space.

Arriving at the reception desk, you confirm your appointment with the nurse and sit in the waiting room. Admiring the space art and pictures of the orbital hotels on the walls, your gaze stumbles upon a few magazines lying on a glass table. "Elvis lives on the Moon!" and "Astonishing results with 0-g cornflakes diet" are some of the headlines that scream at you. You don't bother to read any of the stories and instead choose to study the other people in the room. A happy young couple, maybe planning to spend their honeymoon in space, a middle-aged woman with her teenage daughter and an older man with a walking cane. He will be able to move much more easily without gravity; there will be no need for his cane in the space hotel. After 40 minutes or so your name is called and you enter the doctor's office.

As you sit down she introduces herself and starts to ask some routine questions about your health: "Do you get nervous staying in small rooms? Have you ever experienced any heart problems?" You answer them, while a nurse comes in to collect the plastic bottle with the urine sample you took with you. The sample will be put in an automatic analyzer that checks for any abnormalities that could indicate problems with your health. The doctor checks your blood pressure and heart rate, listens to your breathing and checks your reflexes with a little hammer. No problems so far.

Next she leads you into an adjacent room, asking you to place yourself in a big chair and put your face in a mask in front of you. The mask blocks out any view of the room but through it you are looking at some type of screen on which various colored lights can be seen. The doctor straps you in tightly and the chair starts to rotate. On the screen, you see the lights starting to move around, spiraling down into the distance. You begin to feel a little uneasy, as you lose your orientation and sense of direction. Fortunately, the unpleasantness lasts for only a short period. The device is stopped and the doctor unfastens the straps holding you in the seat. Carefully, you step out, putting a hand on the wall to steady yourself, as you are slightly dizzy from the experience (see Figure 3).

The doctor studies the results of the experiment on a computer screen. "The graphs indicate that you may be sensitive to some mild space motion sickness for the first two days of your trip. Don't worry, about two out of three people

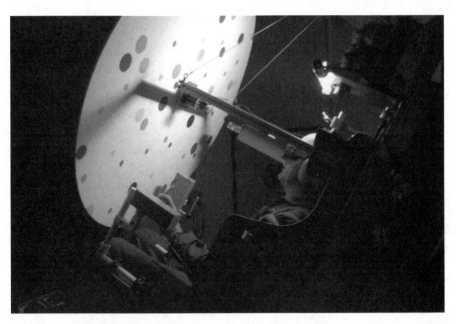

FIGURE 3   On Earth, an astronaut stares at a rotating disk with colorful spots to investigate the space sickness phenomena. In the future, knowledge gained from such experiments may help doctors to determine one's susceptibility to space sickness before flight. [Photo: ESA]

going into orbit have some trouble with it. The flight crew can administer injections with an effective drug if needed; just ask for them if you start to feel sick during the flight. Even so, you may still experience some nausea and dizziness when you arrive into orbit, just don't move around too much and try to keep your head still the first day, okay? And the Promethazine may make you feel a bit slow, but that is a harmless side-effect of the drug, I'm afraid."

The results from the analysis of your urine sample also appear on her screen, indicating a perfectly normal composition. All tests indicate that you are in good health, so she prints out a small report and signs it with a special pen. The ink in the pen contains coded strings of molecules to prevent anyone from falsifying official documents signed by her. She tells you she will send the original report to the training center and wishes you a good flight and a great holiday.

Thanking the doctor, you walk out of her office with a copy of the medical report. You are now medically qualified for your holiday in space.

2

SELECTION FOR SPACEFLIGHT

THINKING of Western astronauts and Russian cosmonauts, you may have visions of athletic, highly educated pilots and scientists, selected from thousands of people aspiring to a job in space. But does it really require supermen and superwomen to go into orbit?

The first people selected to become astronauts were all military jet pilots, male, in good shape, psychologically stable, intelligent and used to making quick decisions in life-or-death situations. Trained to handle major problems and minor disasters without panicking, they were to blaze the trail for humanity into outer space: a super challenge requiring super men.

Before the first people were launched into space, almost nothing was known about what weightlessness would do to the human body. Experiments were performed on rats, dogs and monkeys that were launched to extreme altitudes and even into Earth orbit. The animals handled their flights well, but all they really had to do was breathe. An exception was Ham, the chimpanzee that NASA launched in their new Mercury capsule in 1961, who was required to pull different levers when

15

FIGURE 4   *In 1961 NASA launched the chimpanzee Ham in a Mercury capsule to test whether humans could survive a spaceflight. [Photo: NASA]*

certain lights on his display came on. If he completed a task correctly, he was rewarded by a small banana candy. If he made a mistake, however, he was penalized by a low-voltage electrical shock (see Figure 4).

Ham performed well during his flight, although he was subjected to much higher acceleration forces than anticipated; a faulty valve had caused too much propellant to be injected in the engine, leading the Redstone rocket to climb too quickly and too steeply. When the engine shut down prematurely, because the high propellant consumption emptied the tanks sooner than planned, the emergency escape rocket pulled the capsule from the Redstone. This again put considerable force on its little passenger.

Nevertheless, the capsule retrieval team found Ham to be in quite normal condition after his safe landing and subsequent checks proved that the flight had not caused any problems to the health or mental stability of the chimp. Eventually Ham died of old age in 1983, in a zoo in North Carolina.

However, humans are not chimpanzees. The animals shot into space had no idea that they were sitting on a potential bomb while being launched some hundreds of kilometers above the surface of the Earth. They did not realize that they came back in a fiery reentry that scorched the outsides of their capsules, or that they could spatter all over their home planet's surface if the parachutes malfunctioned. Could a human endure such mental stress? Furthermore, humans were eventually expected to do something useful in space. Would they become too disoriented to perform their work? Would they be so overwhelmed by the Universe that they would slip into some kind of shock or depression? Would astronauts be able to swallow their food without gravity? Or maybe the shape of their eyes would change so that they would be unable to see anything. To be able to cope with all these potential problems, some very special people were required.

The Americans selected test pilots, although there were initially plans to recruit circus artists, stunt men and race car drivers. Test pilots seemed to have all the skills NASA was looking for and were, moreover, used to following orders, something less obvious for other types of daredevils. Russia too selected a group of their top pilots. As a typical example of the communist hero, a young jet pilot of humble farmer origin was to become their first cosmonaut.

Although NASA was aiming to get the best pilots available, the top hot shots who risked their lives flying ever higher and ever faster in increasingly advanced planes were not too interested. Being strapped in a can on top of thousands of liters of explosive propellants was not their idea of a career advancement. Furthermore, the first astronauts were not supposed to really fly their space vehicles, they were merely envisioned to be biological specimens for studying the human body in space. Only later, when more experience could be obtained with life in space, did the scientists' plans allow astronauts to actually pilot capsules and spaceplanes and perform experiments. No wonder many of America's top test pilots originally declined the offer. What NASA finally got were the would-be hot shots: young, excellent pilots who had not made it to the top just yet, and competitive jet jockeys wanting to make their fame doing something new and magnificent.

The astronauts that NASA finally selected for their Mercury crewed space project were not too happy with the fact that they were not going to be much more than intelligent monkeys up there. In fact, Ham the Astrochimp had beat them to the first flight! They put pressure on the engineers and scientists to convert the Mercury capsule into a vehicle with a window and control facilities in addition to the automatic systems. In this way, the astronauts would be able to really command the capsule. In the end, they got what they wanted and were able to fly the capsules themselves. This saved the lives of a few astronauts when automatic systems in their spaceships were lost during flight.

The early Russian cosmonauts were not supposed to do much in their fully automated Vostok capsules; they just had to sit tight and describe what was happening during their orbits around the Earth.

The selection process on both sides in the American–Russian space race was extremely tough. NASA candidates were put on treadmills to measure their stamina and strapped into fast-spinning centrifuges to see if they could stand the accelerations of a rocket. They were locked into dark isolation chambers where they were tortured with sudden bright lights and strange sounds. Every opening in their body was jammed with medical instruments and they were also tested to the limit psychologically, as realistically depicted in *The Right Stuff*.

The selection process for the first Soviet cosmonauts was at least as tough as in the USA. It involved, for instance, being locked up in an isolation chamber for anytime between one and ten days, but candidates were not told beforehand how long the experiment was going to last. Alone, they spent their time in the chamber with intelligence tests and physical exercises while the physicians continuously changed the atmospheric pressure inside. There was no communication with the outside world except with the doctors. In another experiment, a subject had to do a math test while someone was whispering the wrong answers through a headphone. These tests were designed to see if a candidate could stand being alone in an isolated environment and be able to concentrate under high psychological pressure.

Probably the worst experiment would-be cosmonauts had to go through was the oxygen starvation test, in which the candidate had to write his name over and over again on a piece of paper while the oxygen level in the room was being reduced. On a movie recording of one such test, Yuri Gagarin (who eventually became the first human in space) is seen starting to write erratically when the oxygen level becomes too low for his brain to function normally. In the end, the poor cosmonaut drops his pen and paper pad, stares into nothingness for a few seconds, then becomes unconscious. The doctors wanted to check how long a candidate would be able to function in space in case a leak developed in his capsule.

Only men went through the selection process for the first group of NASA astronauts, since there were no women among America's experienced jet pilots. Jackie Cochran, a famous pilot who grew up in poverty but made her fortune in the cosmetics business, wanted to see whether women pilots could perform the same tests as the Mercury astronauts.

Jackie held over 200 flying records and was the first woman to fly faster than sound while she was training for a speed record attempt under the guidance of famous test pilot Chuck Yeager (who himself was the first to break the sound barrier). Being a quite overwhelming personality with many influential friends, she could get anything done. Too old for the demanding experiments herself, Jackie wanted to invite a group of younger women pilots to go through the rigorous astronaut selection process.

For carrying out the tests she contacted the Lovelace Foundation, a research laboratory and medical clinic set up by the uncle of director W. Randolph Lovelace II. Randolph, a physician, pilot and former medical director of NASA, agreed to place his sophisticated facilities at her disposal, while Jackie and her husband Floyd Odlum would pay for most of the expenses. The tests would be performed in secrecy but the results would be sent to NASA for evaluation.

In 1960, 29-year-old Geraldyn "Jerrie" Cobb entered Phase One of the test procedures. Jerrie started flying at the age of 12, held three world speed and altitude records and had accrued over 7,000 flying hours in 64 different (mostly propjet) airplanes. She was subjected to five days of tests of her health, stamina, and tolerance. This meant that Jerrie had her teeth and body X-rayed, her retinas photographed, her blood and urine analyzed, and she had to ride an exercise bike to the absolute limit of her endurance. These were, however, the more pleasant tests; other experiments had her swallow nearly a meter (3.3 feet) of rubber tubing, castor oil, radioactive water, and chalk-like liquid barium. Ice-cold water was injected into her ears to see if she was susceptible to vertigo and the physicians stuck 18 needles in her head to record brainwaves.

As Jerrie's results were excellent, it was decided to send her to another medical laboratory for Phase Two's psychological and psychiatric assessments. There she was locked up in a small, airtight and soundproof isolation tank filled with warm water that was maintained exactly at her body temperature. In complete darkness, she had no feeling of warm or cold and could not hear any sound. Most people would become disoriented, experience hallucinations and start to panic after some time, but when Jerrie was removed from the tank after 10 hours of total isolation, she showed no signs of upcoming hysteria or serious disorientation.

For the final step of the process, she was sent to the Naval Air Station in Pensacola, Florida, for a further two days of psychological and medical tests. She was put in rotating rooms, dumped in a deep swimming pool dressed in a full flight suit and parachute pack, and was subjected to high *g*-forces (accelerations) in an airplane while her brainwaves were being monitored. Again, she withstood the ordeals in an excellent manner.

Impressed with Jerrie's results, Jackie Cochran and the Lovelace Foundation invited more women pilots to participate in the experiments. A selected group of 25 women went through the Phase One and Phase Two tests, 12 of whom were found to be psychologically and medically qualified to train as astronaut candidates. In fact, the women performed some of the tests even better than NASA's Mercury astronaut selection. While NASA's seven astronauts were known as the "Mercury 7," the women would later become known as the "Mercury 13."

NASA didn't invite the women into their astronaut corps, however. Although the "Mercury 13" fought hard for the chance to become astronauts and accused NASA of unreasonable discrimination, Jackie Cochran stated that she agreed with NASA's position:

> The manned space flights are expensive and also urgent in the national interest, and therefore in selecting astronauts it was natural and proper to sift them from the group of male pilots who had already proven by aircraft testing and high-speed precision flying that they were experienced, competent and qualified to meet possible emergencies in a new environment. Because very few individuals will be used as astronauts in the near future and there is no shortage of well-trained and long-experienced male pilots to serve as astronauts, it follows that present use of women, as such, in this connection cannot be based on present need.

Although Russia achieved another space first by sending a woman, Valentina Tereshkova, into orbit in 1963, none of the Mercury 13 ever made it into space. NASA's first women astronauts didn't fly before the arrival of the Space Shuttle in the early 1980s.

After Yuri Gagarin became the first human in space and other Russian cosmonauts and American astronauts followed, it became clear that eyes don't really change shape dramatically in space, eating is quite easy, and sleeping in orbit is not very difficult.

Astronaut John Glenn had a small eye chart taped to his Mercury console, because researchers expected his eyes to swell in microgravity and his vision to blur. Glenn was supposed to look at the chart, and if the letters got blurry he had to do an emergency reentry. No such problems occurred with any of those early astronauts and cosmonauts; the human body proved to be very able to handle microgravity.

Because of this, selection criteria went down a little, but for 20 years space explorers were still required to be in extremely good health and had to be experienced jet pilots.

Nowadays, astronauts are no longer required to be supermen/women and test pilots. Especially with the arrival of the Space Shuttle, more and more scientists, doctors, engineers and even some politicians get the chance to fly into space and do scientific experiments. On its last, ill-fated, flight the Space Shuttle Challenger even had a high school teacher on board as part of NASA's "Citizen in Space" program. However, that proved to be a step too far. The idea behind the "Citizen in Space" program was to fly "ordinary" people with the Shuttle; this was good for public relations and would prove that NASA had opened the space frontier for almost everyone. This dream was violently shattered when the Challenger was ripped apart shortly after launch: a solid rocket booster had been leaking fire from its side, which had burned through one of its attachment structures. When the booster got partly detached, it crashed into the External Tank and caused the break-up of the whole vehicle. None of the astronauts survived.

The Space Shuttle was redesigned and the operations plans critically reviewed and adapted to increase the vehicle's safety and reliability, but it was decided that no non-professional passengers would be flown.

Because of the danger and complexity of spaceflight, it still requires above-average health, a high IQ, and a lot of experience and education to be selected as an astronaut in America, Canada, Europe or Japan or as a cosmonaut in Russia.

The selection for NASA's and the European Space Agency's science astronauts – the people who run the experiments in the Shuttle and the Space Station – still checks for stamina during parabolic flights and in centrifuges. There is also an isolation test where the candidates have to spend some time alone inside a plastic ball. However, much more attention than in the past is given to the scientific knowledge of the would-be astronauts, and not only in their own fields of expertise but also in a broad range of other sciences. As an astronaut, the candidates will have to be able to work on all kinds of different experiments, from crystal growth to taking blood samples and the melting of metals.

Astronauts go into space to work; they do experiments, repair satellites, build space stations, fly spaceships and have to be expert at what they are doing. Time in orbit is limited and expensive, and therefore productivity is very important. Astronaut training takes a lot of time and costs a lot of taxpayers' money too; a not-so-fit astronaut may be grounded because of health problems before his or her training and skills pay off. For these reasons, space agencies still select only very fit, psychological stable and highly educated, relatively young people. Since few astronauts are required at the moment, and as many people aspire to fly into space, there are plenty of candidates to choose from.

Apart from being scientists and pilots, astronauts have always been public symbols too, whether they liked it or not. In particular, the first astronauts were seen as heroic patriots, perfect role models, the best that communism or capitalism had to offer. As such, their private lives were also heavily scrutinized during the selection process; happily married fathers were preferred over womanizing bachelors and divorcees. Being good at public relations and having a nice smile for the cameras was also a plus during the selections.

During their entire career, the early pioneers of crewed spaceflight found themselves constantly in the view of the media. With all the glory and attention of beautiful groupies attracted by their fame, keeping the role of Mr. Perfect sometimes proved difficult. Nevertheless, the astronauts and cosmonauts were expected to maintain an immaculate image.

In 1965, trainee Duane Graveline was kicked out of the NASA astronaut corps because his wife accused him of psychological harassment and wanted to divorce him. The divorce of Apollo astronaut Donn Eisele, who had already lived apart from his wife for some years, was anxiously kept secret.

Three candidate cosmonauts were dismissed in 1963 when the Russian military police found them highly intoxicated on a railway platform. Grigory Nelyubov, previously envisioned as becoming Russia's third man in space, was even transferred to a remote air station in Siberia because he refused to apologize for his behavior. There he tried to convince his fellow pilots that he had once served as backup for Yuri Gagarin, but no one believed him. Deeply depressed, he threw himself under a train in 1966. If you try to find Nelyubov in a picture of the first cosmonaut group, you will probably not succeed: he was airbrushed out by the Soviet government. The only picture to escape the scrutiny of the political information officers was discovered by some keen-eyed Western historians in 1973.

Another cosmonaut was rapidly sent home when he was found in a French pub while he was supposed to be attending the Le Bourget airshow in Paris.

Even Yuri Gagarin, after becoming famous as the world's first spaceman, jumped out of a window to avoid his wife finding him in a hotel room with another woman. He landed badly and had to stay in hospital for a month, but the whole affair was kept secret. The misbehavior of Gagarin was caused by his frustration over the fact that, as a national symbol, he had become too important to be allowed to risk his life on another spaceflight. In his case, fame and good reputation was actually blocking his further career as a cosmonaut.

With the coming of space tourism, the requirements are changing. Dennis Tito and Mark Shuttleworth, the first space tourists who made a flight to the International Space Station with a Russian Soyuz launch vehicle, had to go through physical and psychological evaluations but were not required to be extremely fit scientists or pilots with spotless images. Because they were not going to do any real work in space, only reasonably good health was required of them (see Figure 5) – plus eight months of intensive training and the $20 million they had to pay for the ticket, of course.

Japanese journalist Toyohiro Akiyama, sent up to the Russian Mir space station by television station TBS in 1990, admitted that he was not

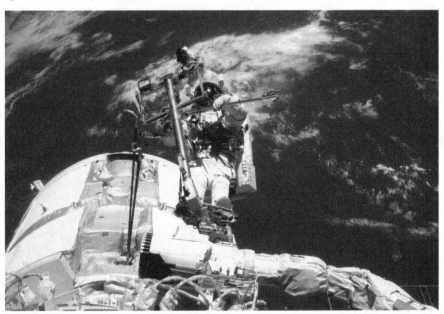

FIGURE 5 *Construction work in space is dangerous, difficult and requires extensive training. Such tasks are the reason why professional astronauts are required to be highly educated, in extreme good health and psychologically exceptionally stable. Space tourists, however, will not need the same capabilities. [Photo: ESA]*

particularly fit and smoked four packets of cigarettes a day. Veteran Mercury astronaut John Glenn made his second flight in space at the age of 77. Although he was in good condition for his age, he was certainly not as physically fit as any other astronaut before him.

Flying in space turns out to be not so difficult at all, as long as you don't have to perform complex operations or guide a Shuttle Orbiter flying at 25 times the speed of sound to a landing without engines. Weightlessness actually requires less strength and stamina than living on Earth! Many handicapped people will find it easier to move around in microgravity, since legs are not really required to move around in space and there is almost no stress on muscles and bones.

The items that are physically most difficult to cope with are the g-forces during launch and return from space. These forces are the result of accelerations and decelerations, which make you feel heavier than you actually are.

One g is your normal weight, caused by gravity. In a fast accelerating sports car, the horizontal force with which you are pushed back into your chair may be 0.7 g, i.e. equivalent to 70 percent of normal gravity. Going up in an elevator, the vertical acceleration added to normal gravity results in forces just a little over 1 g. Likewise, going down in an elevator initially results in a little less than 1 g, making you feel lighter. Once the elevator no longer decelerates or accelerates, but maintains a constant velocity, only the normal 1 g force remains.

NASCAR race drivers can experience up to 3 g during turns, while passengers on the Incredible Hulk ride at Islands of Adventure in Orlando experience up to 4 g, but only for a few seconds.

Fighter pilots regularly pull close to 8 g during high-speed turns. They are taught to strain all their muscles in the lower body during such maneuvers, to keep sufficient blood in the head and thus avoid passing out. Moreover, they wear special "g-suits" with pants that can inflate to prevent blood from flowing into the lower body.

In the past the acceleration during launch or the deceleration of a capsule returning from space into the atmosphere could subject astronauts to 8 g for several minutes. If your normal weight is 70 kilograms (155 pounds), that means that you would have an equivalent weight of 560 kilograms (1,240 pounds)!

High g-forces make it difficult to breathe, as pushing out your "heavier" chest to take in air becomes hard. This is especially true when your body is weakened by a long stay in microgravity.

In a vertical direction g-forces can make your blood pool into your lower body (if accelerating upwards) or flow to your head (if accelerating downwards). In the first case, accelerations higher than about 5 g cause tunnel vision and eventually black-outs. In the other case it can result in so-called "red outs."

To endure such forces for relatively long periods of time, and without passing out, astronauts required special seats shaped to their bodies to spread the weight evenly. They also use g-suits similar to those worn by fighter pilots. The human body is best able to handle g-forces in a horizontal position, which is why astronauts are launched, and usually return from orbit, on their backs. When lying down, the weight can be spread over the entire back, the head can be supported, and blood does not flow away from the brain.

Accelerations on today's crewed launchers are relatively benign. The Russian Soyuz rockets put 4.4 g on the cosmonauts, while the maximum reentry deceleration on board the Soyuz landing capsule is about 4 g. However, in 1975 the crew of Soyuz-18 was subjected to over 21 g when their launcher malfunctioned at high altitude and their emergency escape rockets had to save them from the falling rocket. One of the cosmonauts suffered internal injuries from the extreme g-forces and the rough landing tumbling down a mountainside, and never flew again.

Another hair-raising incident happened in 1983, when the rocket for Soyuz mission T-10 caught fire. Luckily, the crew sitting on top of it were pulled to safety by the powerful emergency escape rockets. For a brief period the cosmonauts were subjected to almost 16 g, but medical examinations afterwards showed no sign of injuries.

A crew returning from the International Space Station in 2003 were subjected to about 9 g when their Soyuz capsule entered the atmosphere more steeply than predicted due to software problems. The force made their tongues roll back in their mouths and they could hardly breathe, but they survived the landing with no apparent lasting effects.

The accelerations experienced during a launch with the Space Shuttle are relatively low, not more than three times the gravitational attraction force of the Earth. That's comparable to the g-forces encountered when going through a loop in a roller coaster, although for a longer period (see Figure 6).

The Shuttle Orbiter's g-forces during return and braking in the atmosphere are even lower, enabling the astronauts on the Orbiter to return in an upright sitting position, like flying in a conventional airplane.

Future launchers for people are envisioned to have similar maximum g-levels as the Space Shuttle, putting only low stresses on the bodies of its passengers. Moreover, early space tourists are likely to remain in orbit for

FIGURE 6   *A Space Shuttle launch looks violent, but does not stress the human body much more than a ride on a roller coaster. From a physical point of view, any healthy person could easily take a flight with a Space Shuttle. [Photo: NASA]*

short periods only and their bodies will therefore not weaken much because of life in microgravity. Any person of reasonable health will easily be able to tolerate the low *g*-forces of a spaceplane returning into the atmosphere.

Other than medical requirements, there may be more criteria you have to face before becoming a space tourist. After NASA lost the battle to prevent the Russians from launching Dennis Tito, an agreement was made with all International Space Station partners on crew selection criteria. If Russia could not be stopped from sending space tourists to the station, at least NASA wanted to prevent problems and embarrassments as much as possible.

The new rules state that any candidate will be judged on medical history, behavioral suitability, language skills, and adherence to the space station "crew code of conduct" by the agency responsible for the flight, whether it be the Russian, American, European, Japanese or Canadian space agencies. Potential space station visitors will have to undergo background checks and assessments on their past and present conduct. They may be disqualified for criminal, dishonest, infamous, or notoriously disgraceful conduct in prior employment or military service, alcohol or drug abuse, or membership of organizations that are considered to be offensive to any of the space station partners.

Tito would easily have passed these criteria had they been applicable for his selection. In fact, the majority of people would have no problem passing these additional background checks.

Space Adventures, the company that arranged the deal between Dennis Tito and the Russians, offers what it calls an Orbital Qualifications Program that includes all medical examinations and tests required to be flight certified by the Russian space agency, Rosaviakosmos. After passing the tests, you are qualified to go on a tourist trip on board a Soyuz spacecraft and visit the International Space Station.

So, if you are now considered to be fit enough to fly as a passenger in an airplane, then medically you would probably be able to be launched into Earth orbit. However, it would all depend on the duration of the trip and the medical facilities available on the flight. People who could, for instance, develop serious heart problems during the flight may have to be excluded as an early return may not be easy and would be very expensive. Nor would pregnant women be allowed on board, owing to the danger of damage to the fetus by the higher levels of radiation above the atmosphere.

In the future, if you want to go on a trip in Earth orbit, you may have to consult a doctor for a brief medical and psychological check. Apart from special cases, it is likely that virtually anyone will be fit enough to travel.

WHEN CAN YOU GO?

Assuming that you do not have tens of millions parked in a savings account to be spent on the holiday of your life – like Tito and Shuttleworth – or cannot find some serious sponsoring, you may have to wait for some time. In the meantime, you could fly to the edge of the atmosphere as a passenger in a Russian Mig-25 jet. For about $12,500, you are able to observe the curvature of the Earth at an altitude of 25 kilometers (16 miles).

It is also possible to spend a holiday living as a cosmonaut in Russia's Star City cosmonaut training center, where you can take rides in the giant centrifuge to feel the acceleration forces of a launch. In a giant swimming pool, you can float around in a spacesuit, weighted just enough to create an equilibrium between the gravity pulling you down and the buoyancy trying to push you up to the surface. For around $10,000, it is possible to hover over a space station mock-up, just like an astronaut in the weightlessness of space. At Star City, you can also experience real microgravity during parabolic flights in a big Ilyushin-76 airplane. While the plane follows a parabolic trajectory with a steep climb and a free fall after going over the top, you float around in the large cabin for about 20 seconds. This is repeated 20 or 30 times per flight and is also available in Russia for about $5,500, including your stay at a luxury hotel. The company Space Adventures has been promoting these holidays via the Internet for some time now.

Several companies all over the world are competing for the $10 million X-Prize – a competition for private enterprises to develop and launch a suborbital vehicle able to carry three passengers to an altitude of at least 100 kilometers (62 miles). The reusability of the rocket has to be proven by making a second flight with the same vehicle within two weeks after the first launch.

The X-Prize vehicles are not designed to go all the way into orbit, but merely to shoot up to high altitude and then fall back. During the unpropelled phase of the ascent and the descent in the higher parts of the atmosphere, there is no acceleration from the engines and hardly any deceleration from the air; the only force working on both the vehicle and the passengers is gravity. Since both accelerate or decelerate equally, the chairs and floor of the vehicle are no longer pushing at the crew, and the people inside the craft will thus experience weightlessness. It is like being inside a falling elevator, but without the hard landing at the end. The principle is the same as for the parabolic flights with the Russian airplane at Star City, but the weightless conditions will last for almost 5 minutes instead of 20 seconds.

Some companies competing for the X-Prize are aiming at simple single-stage rockets, copies of the infamous V-2 rocket of Word War II, or rockets launched from airplanes; others go for winged spaceplanes or even pulse-jet driven flying saucers.

If or when any of these developments will be successful for space tourism is unknown but flights with one of these vehicles are already offered by Space Adventures for $98,000, including a video of the flight and a commemorative medallion. At the time of writing this book, the system that is to make this possible is the Russian-built Cosmopolis XXI, a rocketplane launched from the back of a (already existing) high-altitude jetplane, or the Xerus rocketplane of the California-based XCOR Aerospace company (see Figures 7, 8 and 9). In both cases a professional pilot would fly two passengers to an altitude of 100 kilometers (62 miles),

FIGURE 7 *The Xerus is meant to take off like an airplane and fly space tourists to an altitude of 100 kilometers (62 miles). Passengers will be able to see the curvature of the Earth and be weightless for a few minutes. [XCOR Aerospace, rendering courtesy of Mike Massee/XCOR Aerospace]*

FIGURE 8   *Looking like a conventional airplane, the Xerus will use a rocket engine instead of a jet engine to fly at altitudes where there is little oxygen to burn. [XCOR Aerospace, rendering courtesy of Mike Massee/XCOR Aerospace]*

FIGURE 9   *The Xerus will be the size of a small airplane, but nevertheless able to take two people, not including the pilot, to the edge of space. [XCOR Aerospace, rendering courtesy of Mike Massee/ XCOR Aerospace]*

high enough to experience about 5 minutes of microgravity, look out at the blackness of space, and observe the Earth's curvature.

The Russian project has the advantage that the required rocket engines and the carrier airplane already exist, although the rocket engine for the American vehicle has already been tested.

So far, over a hundred companies and individuals have signed up and made cash deposits for a ride with Space Adventures. For companies such as First USA Bank and Pizza Hut suborbital flights offer great opportunities for promotions, such as contests with a space ride as first prize. One of the individuals who booked a flight with Space Adventures is 63-year-old Wally Funk, one of the 13 women who trained to be Mercury astronauts in the early 1960s. At long last, she hopes to get an opportunity to fly into space as a paying passenger.

However, Scaled Composites, the Californian company of famous aircraft designer Burt Rutan, is far ahead of the competition in the race for the X-Prize. Rutan's design involves a three-seat mini-spaceplane named SpaceShipOne, which is initially attached to a larger twin-engined turbojet aircraft. On July 21, 2004, the White Knight carrier plane brought the small SpaceShipOne with 62-year-old pilot Mike Melvill to an altitude of 14 kilometers (8.7 miles). The spaceplane was dropped into a gliding flight, then fired its rocket motor for 80 seconds, steeply climbing and achieving a maximum speed of no less than 4,000 kilometers per hour (2,500 miles per hour). After rocket engine cut-off, the plane continued upward, unpowered, to an altitude of more than 100 kilometers (62 miles). This coasting phase, and the following free-fall back to Earth, were under microgravity conditions and lasted for about 3 minutes. During its fall, the plane's two tailbooms and back parts of the wings were put in a vertical position to achieve a high-drag configuration that allowed a safe, stable return in the atmosphere. The increasingly thicker air decelerated the spaceplane, after which the craft converted back into a conventional glider and gently flew down from an altitude of 24 kilometers (80,000 feet) to the runway in the Mojave Desert in California (see Figure 10).

With his flight Melvill earned his astronaut wings, and became the first private civilian to fly an aircraft into space. Moreover, he became the first person to leave the atmosphere in a non–government-sponsored vehicle. "It was a mindblowing experience, it really was – absolutely an awesome thing," he said after landing.

To win the X-Prize the team still has to make two such flights within two weeks. At the time of writing this is likely to happen before the end of 2004.

Rutan's team has already developed many successful, innovative aircraft

FIGURE 10   *Launched from under another airplane, SpaceShipOne is designed to launch three people to an altitude of 100 kilometers (62 miles) on a regular basis. [Photo: Scaled Composites]*

over the years. They gained worldwide fame with their Voyager aircraft, the only plane ever to fly around the Earth without refueling or landing. Work on the SpaceShipOne/White Knight concept began in 1996, but the secret full-development program only started in April 2001, hidden from the public and the competition by the inhospitable Mojave Desert. It was therefore a great surprise, and shock for the competition, when in April 2003 Rutan's team presented a fully operational White Knight high-altitude plane, a mobile mission control center, a mobile propulsion test facility, and a prototype of SpaceShipOne.

To finance the project the company got a $30 million grant from Paul Allen, Microsoft cofounder and the third wealthiest person in America. $30 million is considerably more than the $10 million the team would get from winning the X-Prize, but Allen is in it for the adventure, not for the money. Allen's and Rutan's objective is not merely to win; primarily, they want to jump-start a renaissance in the human spaceflight industry and put the adventure back into the overly bureaucratic aerospace business. Once mature, they plan to sell the technology to a large industry or space tourism company. Rutan estimates that, within 10 to 15 years, the price of a ticket for a seat on board a fully operational suborbital spaceplane will be only a few thousand dollars. "Spaceflight is not only for governments to do," he said. "Clearly, there's an enormous pent-up hunger to fly into space and not just dream about it."

Another promising concept is the Thunderbird of the British Starchaser

Industries, a company that has extensive experience on suborbital sounding rockets. Their concept is based on a conventional liquid propellant single-stage rocket with a three-person capsule on top. After reaching an altitude of over 100 kilometers (62 miles), the capsule separates from the booster. Both the rocket and the capsule land by parachute and are reusable (see Figures 11, 12 and 13).

On their website, the company advertises seats on board the Thunderbird for $650,000. They suggest it as a birthday present – for your mother-in-law, for instance.

The X-Prize vehicles are not real spacecraft, as they are not capable of achieving sufficient velocity to go into orbit. They are actually high-altitude rocket vehicles, which makes them smaller, simpler and potentially much cheaper to operate than an orbital launcher such as a Soyuz rocket or the Space Shuttle. Real orbital space tourism is, for now, only possible by using existing launch vehicles and spacecraft, which are very expensive. (Dennis Tito and Mark Shuttleworth each spent an estimated $20 million for their Soyuz launch and stay on board the International Space Station.)

FIGURE 11 *The Thunderbird capsule is meant to accommodate three space tourists. [Picture courtesy of Starchaser Industries Ltd]*

FIGURE 12   The Thunderbird from the UK will resemble a conventional launcher and capsule system, but be completely reusable. [Picture courtesy of Starchaser Industries Ltd]

For more affordable tourist flights into Earth orbit you will have to be a little more patient. It is difficult to say when mass space tourism could start as a viable business; many predictions have been made before and proven to be completely wrong.

During a break in the television transmissions of the Apollo 8 mission to the Moon in 1968, the president of Pan Am called TV station ABC-TV to announce that his airline would begin taking reservations for future trips to the Moon. Rival airline TWA also started to accept reservations for Moon trips. At that time, even space expert Wernher von Braun was convinced that lunar flights would be available to the general public before the year 2000. When Pan Am closed its reservation list in 1971, it contained the names of some 93,000 people, including future president Ronald Reagan.

In 1985 an American company, Society Expeditions, offered orbital flights on board a vehicle called Phoenix E, for $50,000 per passenger. They took 250 deposits of $7,000 (of which $5,000 could be withdrawn at any time, but $2,000 was spent on financing the organization) but the

FIGURE 13   On the launch pad, a Thunderbird rocket stands ready to launch a group of space tourists. [Picture courtesy of Starchaser Industries Ltd]

vehicle was never build. Confidence in the rocket technology develop-ments in the 1980s made some people think that orbital space tourism would become a reality within a few years. It turned out not to be so.

Space tourism will start with the happy rich few taking the first tourist steps into space, millionaires bank-rolling themselves into orbit. Subse-quently, space trips will become cheaper, more comfortable and available to more and more people. This evolutionary process has already started with the first non-professional astronauts paying for a trip to the International Space Station. Further progress depends very much on the developments in reusable launcher technology.

So, assuming you have no more than a couple of thousand dollars to spare for a trip to Earth orbit, when can you start packing?

NASA has slowed down its quest for large reusable launch vehicles. It currently plans to use the remaining three Space Shuttles until 2010, after which a new system will take over the task of transporting astronauts. However, this new vehicle may turn out to be an expendable capsule-type spacecraft rather than a reusable spaceplane.

NASA is planning to have its true second generation of reusable launchers operational around 2020 (the Space Shuttle is considered a first-generation vehicle). The goal is to limit the cost per launch to around $6,000 per kilogram payload, much less than the current $10,000 to $20,000 per kilogram to low Earth orbit but still far too expensive for a viable space tourism industry. The new launch vehicle is also to be 20 times safer than the Space Shuttle with its current statistic 1 in 250 chance of a catastrophic failure. (In reality, with the demise of Challenger and the Columbia, the Shuttle has experienced 2 disasters within 113 flights.)

Access to space for less than $50 per kilogram, and with it the possibility for mass space tourism, is hoped to arrive with the fourth generation of reusable launchers, envisioned to be operational around 2040. This generation of vehicles will be airplane-like, airbreathing vehicles based on new super-light materials, will have efficient rocket engines, and will require no more maintenance than a modern airliner.

Prior to that, the third-generation vehicles would come into use around 2025. Although flights with the third-generation launch vehicles will be quite expensive, the prices could be low enough to make flights possible for the first adventurous individuals with an average income willing to save some months or a year of their salary for their dream flight. Maybe you will be lucky and win a ticket to orbit on board a third-generation spaceplane in a game show, lottery or as part of a promotion for your favorite brand of corn flakes!

In Japan, plans for future crewed spacecraft seriously take into account possibilities and requirements for space tourism. The egg-shaped *Kankoh-maru* of the Japanese Rocket Society, for instance, is a study concept for a fully reusable single-stage spacecraft. It is not a spaceplane as those envisioned by NASA, but a giant reusable rocket without wings, that would take off and land vertically. Specifically designed for space tourism, it would be able to carry 50 passengers into orbit. The vehicle would weigh about 550,000 kilograms (1,213,000 pounds) at launch, be constructed from mainly lightweight aluminum and composite materials and use 12 rocket engines operating on liquid hydrogen and oxygen. The 18-meter (59-foot) diameter, 22-meter (72-foot) high ship is to take off and land vertically from a specially developed spaceport. The craft is envisioned to have a cockpit for a pilot and a flight engineer, while a co-pilot, a navigator and a ground crew chief at the spaceport would be in contact with them via a satellite link. The passengers would be positioned on top, in a circle around the outer hull of the vehicle. Each would have a window straight in front of his or her face, guaranteeing an easy and unobstructed view for the duration of the flight.

It seems to be almost certain that space tourism will grow into a viable business and a major economical factor as the limited market surveys done so far indicate that many people are interested in going into space. For instance, one survey concluded that about 70 percent of the Japanese want to make such a trip and that most of them are willing to pay at least three months of salary to make a flight. A recent market study by the Futron Corporation indicates that commercial space tourism could generate over $1 billion revenues by 2021. A good promotional campaign, with advertisements and publicity stunts like flying famous people into space, should increase the popularity of space tourism.

The market is there, and although the technology is not yet available it is under development. It's only a question of time before your spaceplane will be ready for boarding.

SAFETY FIRST

Even today being an astronaut is still not a very safe occupation; launchers have a disturbing tendency to blow up, space is without air but full of deadly radiation, and to come back to Earth you have to smack your spacecraft into the atmosphere at 25 times the speed of sound.

Until now the United States has experienced two fatal accidents during flight, both with the Space Shuttle: the Challenger launch disaster in 1986 and the break-up of Columbia during reentry in 2003. Russia lost one cosmonaut and his capsule because of a malfunctioning parachute system in 1967, and a crew of three due to a leak in the pressure cabin during reentry in 1971.

Some other accidents during flights could easily have resulted in fatalities. The crew of Apollo 13 barely made it back home alive after an oxygen tank explosion on their way to the Moon in 1970. In 1975 the crew of Soyuz-18 were saved from their falling rocket, already at 192 kilometers (119 miles) altitude, by their emergency escape rockets. Also in 1983 the cosmonauts on board Soyuz T-10 narrowly escaped death, when emergency escape rockets pulled their spacecraft away from the rocket that was exploding on the launch pad.

With a total of about 240 human space missions to date, the average of fatal accidents is 1 per 60 flights. This is extremely high when compared to commercial aviation: taking a plane to go on holidays exposes you to only a 1 in 2 million probability of not arriving at your destination alive. Even parachuting is very safe in comparison, with only about one fatal accident per 100,000 jumps.

Mass space tourism will probably require a safety level close to that of today's airliners, otherwise the market will be too small. Not many people will be comfortable making a spaceflight that is more than 10,000 times more dangerous than flying on an airliner. Moreover, the insurance fees would be prohibitively high.

The current launch vehicles are risky because to get into space with today's technology requires us to push these machines to their limits. Rocket engines operate at extremely high pressures to generate sufficient thrust, which in turn puts heavy loads on the launcher's structure. This would not be so difficult to cope with if launchers could be made heavy and strong, like tanks. However, to carry a useful payload into orbit, launchers need to be very light and are consequently rather fragile. For instance, the Atlas rocket that launched the first American astronauts into orbit had such a thin metal skin that without internal pressure it would have crumpled under its own weight. It was like a soda can, which can handle quite a load when full, but can easily be crushed when empty.

Also, to keep the weight to a minimum, launchers seldom benefit from the same amount of backup systems as a large airplane. When an engine on a launcher fails at an early stage in the flight, the mission is lost. At best the crew can escape the inevitable crash by using some kind of escape system, but turning around and flying back to base like a malfunctioning airplane is not possible (see Figure 14).

Launching also involves huge amounts of extremely dangerous, explosive propellants. The Space Shuttle's Solid Rocket Boosters are notoriously unsafe. They resemble huge firecrackers, using a core of solid propellant rather than tanks with liquid fuel and oxidizer. A crack in the solid propellant grain in these boosters could result in too-rapid burning, overpressure, and a devastating explosion. There is nothing anyone could do to prevent such a disaster once the boosters have been ignited, because, like fireworks, they cannot be extinguished.

The fact that the Shuttle's Solid Rocket Boosters, unlike liquid propellant systems, cannot be throttled or shut down also means that the Space Shuttle has no flight-abort possibilities while they are in operation. If one or more of the liquid-propelled main engines fail, the Shuttle can fly back to its launch site, land elsewhere in the world, or abort to a lower orbit than intended, but only after the two rocket boosters have burned out. Before that, there is no escape.

Another risk-increasing point is that due to their expandable nature, fragility and high cost, modern launchers cannot be test flown to any great

FIGURE 14   Conventional launchers like Europe's Ariane 5 carry huge amounts of dangerous liquid propellants, explosive solid propellant boosters and allow little margin for error. Such a system has no hope of ever attaining airline safety levels. [Photo: ESA]

extent. A new type of rocket is usually declared operational after only one or two test flights, while, typically, 1,000 test flights are made with one single plane before a new airliner goes into service.

Prototypes of new airplanes can be gradually pushed to higher altitudes and velocities during the test phase, but for launchers this is generally not possible. Rockets either stand still on the launch pad, or fly all the way into space; there are no possibilities in between. The first Space Shuttle flight of the complete system also had to be a full mission, with a standard launch, orbit, reentry and a crew on board. Before Space Shuttle Columbia's first flight, the United States had never launched such a large spaceplane or flown one back from orbit. Everything had to operate correctly first time (and fortunately it did).

Individual rockets cannot be test flown at all; you build them and you launch them. You cannot take a launcher fresh off the production line for a test flight before using it on a real mission. The reliability and safety of a launcher mostly depend on what is learned from earlier launches, robust quality control and inspection procedures. Equipment can be tested while the launcher is still on the ground, but how the systems will operate during flight cannot be guaranteed. In contrast, individual commercial planes coming from the factory usually make a few test flights to ensure that everything operates as expected before being handed over to the customer.

Because of all this, future space tourists are likely to fly on board aircraft-like, reusable spaceplanes rather than expendable rockets. During step-wise test campaigns similar to those of commercial or military aircraft, spaceplanes could be tested much more thoroughly than ordinary, expendable rockets. They could have higher safety margins and more backup systems, and enable safe aborts during the entire launch phase. The Space Shuttle, with its various abort scenarios, is partly a spaceplane, but still depends on expendable rocket technology (see Figure 15).

The trick will be to combine the relative simplicity of expendable rockets with the reliability and safety of airplanes, to end up with a safe, economic launch vehicle. However, if they are not well designed, spaceplanes could end up combining the unreliability of expendable launchers with the complexity of sophisticated airplanes.

Once in orbit, danger is less immediate, easier to mitigate but nevertheless always there. The most serious space hazards are impacts of meteors and space debris, and onboard fires.

Normally, meteors are rare in Earth orbit and the chance of one hitting a spacecraft is small. Sometimes, however, our planet moves through a cloud of these chunks of metal and stone, which causes spectacular

FIGURE 15   *Another Shuttle Orbiter safely returns a crew of astronauts after a completing a hazardous space mission. The chances of something going seriously wrong during a Shuttle flight are, however, very high compared to regular airplane flights. Future space tourist vehicles must be made safer. [Photo: NASA]*

displays of fireworks as the meteors burn up in the atmosphere, but also increases the chance of collision with a spacecraft. In the past, satellites have been destroyed by meteor storms. Luckily, the orbits of the meteor swarms around the Sun are well charted, so crewed spacecraft can be kept on the ground if one is due.

Meteors are natural, but space debris can also consist of pieces of broken-up satellites, flakes of spacecraft paint and garbage thrown out from space stations. There is even a glove flying around somewhere, lost by an astronaut when he opened the hatch for a spacewalk.

Small pieces of debris of only a few millimeters can have a devastating impact energy because of the high orbital velocities involved: a tiny flake of steel of 0.05 gram (0.002 ounce) moving in the opposite direction can hit a spacecraft with the same amount of kinetic energy as a 5-kilogram (11-pound) brick at 90 kilometers per hour (56 miles per hour).

The shielding on most spacecraft and space stations can stop small pieces of debris up to 1 centimeter (0.4 inch) in diameter. Big chunks of junk larger than 10 centimeters (4 inches), like empty rocket stages and dead satellites, are tracked by radar from the ground. Their orbits are therefore well known, so satellites and crewed spacecraft can be placed in safe orbits to avoid collisions.

The truly dangerous pieces are those between 1 and 10 centimeters (0.4 and 4 inches) in diameter, since they are able to pierce through a spacecraft's outer hull but are too small to be tracked. The risk of encountering one in low Earth orbit is, however, very low, since most debris there is slowed down by air resistance and burns up in the atmosphere after a few years in orbit.

Most modern low-orbiting satellites are designed to use their last amount of propellant to brake and fall back to Earth, leaving a clean orbit behind. Rocket stages now also vent their remaining propellant to avoid explosions in space, and the creation of thousands of small, untrackable pieces.

No crewed spacecraft or space station has ever been seriously damaged by meteors or debris, although their windows and solar panels show many tiny impact craters after some time in orbit. Future space tourism vehicles will have to be put in clean orbits and be covered by sufficient shielding to reach a sufficient level of in-orbit safety.

Fires also create a very dangerous hazard in space. On board the Russian Mir space station an oxygen-generating candle once caused a serious fire, filling the station with smoke. Luckily, the astronauts were able to extinguish it and save the station.

Space tourism vehicles and space hotels will require sophisticated fire detection and extinguishing systems, and open fires and smoking will surely be forbidden.

One easy way to put out a fire in space is to open the hatch and decompress the spacecraft. Of course, this is only an option if everyone on board wears a spacesuit or the part that is on fire can be closed off from the rest of the spacecraft and vented individually. This last possibility may be used in large space hotels; having everyone wear spacesuits all of the time is only practical during short flights or transfers to a space station.

RADIATION

Another important in-orbit danger, one that is not immediate but slowly creeps up on you, is radiation. The constant radiation levels in low Earth orbit are much higher than on the ground or in an airplane, because not only is the cosmic and solar radiation in space not filtered by the atmosphere, but astronauts are also bombarded by charged particles from the Sun that got trapped in the magnetosphere of the Earth.

High-energy protons and atomic nuclei, gamma rays and X-rays can damage cells in your body by ionizing molecules inside them. Normally this can be fixed but if too much radiation is absorbed the damage can overcome the body's ability to repair it. In this case the affected cells may die, or damage to their DNA may turn them cancerous.

Radiation is everywhere, also on Earth. It not only comes from space but also out of the ground. The average dose of radiation you pick up on the surface of our planet is about 2.6 milli-Sieverts per year (a Sievert is a measurement for the equivalent radiation dose a person is exposed to, taking into account not just the amount of radiation but the amount that is absorbed by different parts of the body). However, radon gas coming from the ground or outgassed from concrete in new buildings can result in levels that are five times higher.

Astronauts in low Earth orbit pick up about the same amount of radiation in a week that people are normally exposed to during a year on Earth. Spacecraft can be shielded from this normal background radiation by dense materials, such as metals but also by water tanks. However, only a limited amount of shielding can be applied otherwise spacecraft would be too heavy to launch. Until huge space hotels with thick radiation shields become available, the length of time space tourists will be allowed to spend in orbit will probably have to be limited.

People working in nuclear plants are allowed annual doses some 10 times higher than the normal exposure on Earth. Using this as a standard for space tourism would result in a maximum total spaceflight duration of 10 weeks per year. This should fit with nearly anyone's annual holiday schedule, but for the staff of space tourism operators it would be a severe limitation. Probably spaceplane pilots and other space tourism workers would be required to rotate between space and Earth-based jobs, working most of the year on the ground. Total lifetime radiation dose restrictions may also result in limited in-orbit career durations.

Everything described above only concerns normal cosmic and solar radiation. However, the Sun sometimes expels sudden bursts of radiation during so-called solar flares, solar particle events or coronal mass ejections. These phenomena can result in doses of thousands of milli-Sieverts, up to 20,000, which can easily be fatal.

In 1972, for instance, a solar particle event sent out so much radiation that astronauts on their way to the Moon would have been exposed to a lethal total dose of 11,300 milli-Sieverts. The event took place between the Apollo 16 and 17 lunar missions, so fortunately no one was en route to the Moon at that moment.

In low Earth orbit the effect of major solar radiation events is dampened to a great extent by the Earth's magnetic field, our natural defense mechanism. Moreover, we are becoming ever better in predicting the Sun's behavior. If a dangerous event is expected, there is enough time for a spacecraft in orbit to return to Earth. No astronaut has ever been caught in a serious solar radiation outburst.

Staying in space and hiding behind heavy shielding is also a possibility. Future space hotels will probably have radiation shelters where guests can hide while a solar radiation event blazes. They last no more than a few days at maximum, thus posing no more than a short disruption to an orbital holiday.

BACK IN THE ATMOSPHERE

After launch, returning from space is the most dangerous phase of a spaceflight. During reentry, spacecraft have to slow down from orbital velocities of 7.5 kilometers per second (4.7 miles per second) or more to normal airplane speed mostly by using aerodynamic friction. As it slams into the atmosphere, the skin of returning capsules and spaceplanes heats up quickly to extreme temperatures. The temperature of the Shuttle Orbiter's nose and front of the wings, for instance, can register 1,600 degrees Celsius (3,000 degrees Fahrenheit).

Protection against this heat consists of either an ablative or a reusable heatshield. Ablative heatshields are commonly used on capsules. They protect the spacecraft by slowly burning away like a log of wood in a fireplace; the outside is very hot, but the flowing gasses take the heat away so the inside stays relatively cool. Ablative shields are very robust and are sure to provide sufficient protection as long as you make them thick enough.

However, ablative shields can be used only once. They cannot be applied for reusable spacecraft like the Shuttle Orbiter, which requires a reusable heatshield. The Orbiter's protection consists of a carbon-carbon nose and wing leading edges where the temperatures are exceeding 1,300 degrees Celsius (2,300 degrees Fahrenheit), a layer of several types of ceramic tiles where the heat is less severe, and insulating blankets where the temperatures are relatively mild. Most of the protection, i.e. the Orbiter's belly, wings and sides, are covered with tiles. These tiles are amazingly efficient in releasing heat; you can hold an uncoated tile by its edges with an ungloved hand only seconds after taking it from an oven, while its interior still glows red hot.

However, the Shuttle Orbiter's heatshield is notoriously fragile; the tiles are so brittle that they would not survive flying through a hailstorm, and even rain can cause severe damage. The impact of an icy piece of External Tank insulation foam on a wing leading edge during launch was responsible for damaging the thermal protection system of the Space Shuttle Columbia. During reentry in the atmosphere, hot gasses entered the wing through a hole created by the impact and caused the loss of control. Columbia fell apart, highlighting once again the dangerous nature of current human spaceflight.

More modern spaceplanes are envisioned to rely on more robust reusable heatshields, probably made of metallic plates. These should make it possible for future spacecraft to return from orbit even in bad weather. They will increase safety during reentry and also make faster emergency returns possible; currently, the Shuttle Orbiter sometimes has to wait an extra day in orbit for the weather to improve at the landing site.

LANDING

Surviving the heat of reentry does not end the danger, because the spacecraft still has to land.

Crewed capsules have always used parachute systems for the final descent stages. Parachutes are relatively simple and reliable systems that do not need active steering or sophisticated spacecraft aerodynamics. However, if a spacecraft comes tumbling down, as happened with Komarov's Soyuz-1, deployed parachutes can become twisted and tangled. This resulted in cosmonaut Komarov's death in 1967.

Capsules and parachutes can also lead astronauts to come down far from their intentional landing site, resulting in possible safety hazards. Crews may land in deserted forests full of wolves (as happened to the Russian Voschod-2 crew) or splash down in the wrong part of the ocean, far away from the nearest rescue ship. Astronauts sometimes had to wait for hours or even days to be found by their rescue teams. Because of these problems landings at night are not advisable with a capsule system. In contrast, airplanes as well as the Shuttle Orbiter regularly make night landings.

Spaceplanes that can actively control their flight like aircraft are an improvement over the old-fashioned parachute and capsule approach.

However, flying back from space is a lot more difficult than simply gliding down under a parachute and requires sophisticated control systems.

A safety hazard related to airplane-like landings is that spaceplanes cannot carry sufficient amounts of propellant for powered landings. For instance, the Shuttle Orbiter comes down all the way from orbit like a glider, without using its powerful engines.

The problem is that with such unpowered vehicles there is only one shot at a successful landing; no possibility of going full throttle and turn around for another try. Future spaceplanes may have small jet engines to assist them during landing, or sophisticated air-breathing engines that can be used both during launch and landing.

Final contact with the Earth's surface is usually quite brutal for a capsule descending under parachutes. In the USA, the problem was solved by landing in the sea where the water provided a relatively cushioned impact. Russian capsules land on solid ground, requiring a set of solid propellant rockets to brake the capsule just before landing.

These landings are also safety hazards, because descending too quickly or having a malfunctioning rocket braking system can result in too hard landings and injuries or worse. Spacecraft landing like airplanes, such as the Shuttle Orbiter, do not need to rely on cushioning or braking systems. However, they need to land on suitable runways, while capsules can come down in one piece virtually anywhere on Earth.

Future spaceplanes should promise dramatic increases in spaceflight safety levels for launch, reentry, and landing. They should constantly be under full control, be able to come back from space at almost any time (day and night, even in bad weather), need not rely on rescue teams for recovery and not need parachutes that might not deploy correctly. Moreover, spaceplanes should offer safe abort possibilities during almost all flight phases. Just as airliners are generally safer than balloons, parachutes, and fighter aircraft, true spaceplanes could also improve safety levels for human spaceflight. Advanced spaceplanes should be designed to be just as safe as modern airliners, thus making mass space tourism a possibility.

3

GETTING READY

FROM the bus you first see the metallic cylinders of the Orbital Destinations rocket museum park blinking in the sunlight, pointing toward the blue open sky above Cape Canaveral, Florida. In the distance the huge Vehicle Assembly Building of NASA's Kennedy Space Center can be seen, where the old Apollo moonrockets and the Space Shuttles were assembled before being transported to the launch pads. However, that is not your destination today.

After a few minutes the bus arrives at the gate, where a large sign welcomes you at the Gagarin Space Camp. Passing that, you drive on to a square overseen from the center by a large white statue. You recognize the figure in the middle as the famous Russian cosmonaut after whom the site is named: Yuri Gagarin, the first man in space. He proved that the human body could withstand a launch and adapt to microgravity without much trouble. And although he was most certainly not a tourist, he was the first to enjoy microgravity and the awesome view of the Earth from orbit.

At the sides of the circular road around the square are souvenir shops, restaurants, virtual reality game booths and cinemas. It is busy in the plaza and the streets connecting to it, as it is late afternoon and most trainee classes have now finished.

The people here are not only tourist-trainees getting ready for a flight, but many are day-trippers visiting the museum and participating in some of the training activities. The short visit programs and attractions not only increase the exposure of tourist space travel to a larger audience but they also encourage the day visitors to take a flight some day.

The bus honks at some pedestrians in the road and stops in front of a large glass building that seems to be carved out of gemstone. It is no doubt the Vostok Hotel, the place you will be staying for the coming days as you prepare yourself for your first space trip. It is named after the spacecraft in which Yuri Gagarin made his flight.

Arriving in your hotel room you find a thick pack of information on the bed, together with your trainee badge, flight suit and helmet. The info pack contains a welcome letter explaining some of the general procedures in the training center, a map of all the facilities, a list with important phone numbers, and a course book with all the information you will need during the classes that start tomorrow. The map shows the location of your hotel and the different classrooms, medical facilities, centrifuge building, neutral buoyancy pool, water landing survival training pool, and the full-size mock-up of the Orbital Destinations spaceplane.

Also included is a waiver that you have to sign to declare that you understand the risks, however small, of participating in the training and the actual space flight. Also, you will agree not to hold the Orbital Destinations organization responsible for anything that could happen to you, other than problems caused by the organization's pure negligence. As customers are informed about this beforehand, you have already taken out special insurance to cover any misfortunes and medical expenses.

The hotel room is quite standard, except for one thing: the toilet. In space, going to the toilet is a more complicated procedure than it is on Earth. Things don't automatically fall down in microgravity, and flushing with water wouldn't work either. Space toilets therefore depend on creating an underpressure to remove urine and feces, like a vacuum cleaner but with much less force.

To create the necessary pressure difference, you have to position your bottom on the toilet just right, use straps to prevent yourself from floating away and make use of a personal plastic adapter to attach to the toilet for urinating. You activate the toilet with a lever to start the fan that creates the slight suction. Paper can also be put in the toilet, but to remove it you must close the lid and pull the lever to ensure that the stuff gets sucked away. In orbit, the solid contents in the toilet's waste tanks are exposed to the vacuum of space after use, causing instant drying as the liquid sublimates immediately into vapor. Urine is collected in a separate tank and periodically dumped into space.

Although simple to operate, the microgravity toilet requires some getting used to. By putting a similar one in your hotel room at the training camp (with a short manual attached) you can practice in privacy when you feel nature calling.

In the early days, crewed capsules did not have proper toilets. Astronauts had no choice but to put their feces, together with some neutralizing chemicals, in the now empty plastic bags from which they had taken their food (which is where the material originally came from anyway). But since the development of the Space Shuttle and the Russian Mir space station, going to the toilet is no longer an embarrassing and complicated procedure.

The first official activity begins that same evening: a welcome and cocktail party to get to know your "crew," consisting of the two spaceplane pilots and the other 31 space tourists who will accompany you into orbit. You find that the people you meet at the party are from all walks of life and range in age from 16 to 80 or so. There are some yuppie types, some young couples, a group of obvious space buffs already wearing their flight suits, and a television crew making a documentary about the experience. What all these people have in common is a wish for some adventure, a desire to enter the mystic realm of space, to experience weightlessness and to enjoy the marvelous view of Earth from an altitude of 250 kilometers (155 miles).

Although space travel to low Earth orbit is now becoming quite routine, it still requires some serious training, of a relatively short duration, to be prepared for spaceflight. The risk of anything going wrong is higher than when flying a typical passenger airplane, therefore regulations do not allow children to enter the spaceplanes, nor people who could have problems during the training, actual flight or emergency procedures owing to their health or disabilities.

SAFETY

The risks of the trip you are about to take are explained in the first class you attend the next day. The most dangerous part of the flight is not the time spent in space, so you are told, but the actual launch and landing phases.

During launch, the loss of one or more engines early in the flight may mean that the spaceplane cannot achieve sufficient velocity to reach orbit. Depending on what phase of the flight you are in at the time of mishap, the pilot's options are to fly back directly to the launch site, to make an emergency landing at any large airport in a broad area along the trajectory, or to go suborbital and land at the spaceport after only one partial orbit. If the vehicle does make it to orbit but reaches a lower than foreseen altitude, you may return earlier than planned; the few atmospheric molecules wandering around in low Earth orbit can slow a spacecraft sufficiently to make it fall back after circling the Earth only a few times. Also, the spaceplane may not be able to dock with the space hotel if there is not enough propellant left to increase the orbital altitude.

More serious problems, such as a fire on board, may make an emergency landing necessary, for which you will receive emergency evacuation and water landing survival training later in the course.

There have been suggestions of equipping each passenger with an ejection seat, but mass and structural constraints make that unfeasible. Furthermore, ejection seats can only be used for a very limited period after take-off, because the spaceplane is soon flying too fast and too high for such a system to be of any use. A better option would be to design the passenger cabin as a separate module that can be ejected from the spaceplane in one piece. This system was proposed for the cancelled European Hermes shuttle but would be so heavy with all its ejection mechanisms, aerodynamic structures and parachute packs that the number of passengers in a spaceplane would be severely constrained.

In orbit, the biggest danger is fire. It eats up oxygen, pollutes the air with smoke and poisonous gasses, and produces intense heat that can kill people and damage vital systems. There is very limited room and air in the passengers' cabin and nowhere to escape to. Fortunately, fires in space usually do not burn as fiercely as on Earth because there can be no convection in microgravity; hot air with a lower density does not rise to make room for new oxygen to feed the fire. Instead of a bright spiky flame, a candle burning in space only develops a small, dim blue ball. Fires in space can only get their oxygen by diffusion (the slow drift of the molecules through the air), which results in a more efficient combustion (hence the dim blue light) but a much smaller flame. Violent fires are only possible if there is a strong airflow from, for instance, a ventilation fan or if the burning material contains its own oxygen (like solid rocket propellant).

On the other hand, microgravity poses problems with conventional means of fire extinguishing: water would not cover the fire but would instead form droplets that float all over the place and perhaps damage electronics; foam and powders would create an enormous mess; while CO_2 gas would suffocate everyone.

Instead, on board the spaceplane fires are extinguished by an aerosol. The automatic fire-extinguishing system sprays out a cloud of small droplets that chemically bind the oxygen in the atmosphere close to the flames. This suffocates the combustion because it can no longer use this oxygen. Moreover, the reaction with the aerosol requires heat that is taken from the fire, thereby cooling it.

The system puts out a fire within seconds and the only thing leftover is a small amount of fine powder that can easily be filtered from the cabin atmosphere.

Another hazard is getting hit by meteors or pieces of space debris. This can cause decompression of the passenger cabin, puncture propellant tanks, or damage other vital equipment. However, spacecraft can be shielded against most of the small particles and can avoid the large ones that show up on the radar. None of the Orbital Destinations spacecraft has ever been seriously damaged by orbital debris or meteorites.

Should a spaceplane have problems in orbit and be unable to return to Earth, then another vehicle can be launched with an aerodynamically shaped emergency docking module attached on top. With this device, any spaceplane in the fleet can dock with any other to take over and bring home the passengers and crew of the crippled vehicle.

During reentry into the atmosphere and the approach and landing phases, the chances of anything going seriously wrong are somewhat smaller than for the launch, since the spaceplane will hardly contain any propellants by then. The vehicle will also be relatively passive as it is designed to glide back from space without engines. Nevertheless, the temperatures and aerodynamic pressures encountered when reentering the atmosphere are tremendous, and the Columbia disaster showed how they can tear a damaged spacecraft apart.

Landing without thrust means that only one landing attempt can be made, since the plane will not have the power to fly around for a second try. This procedure has, however, been perfected since the first unpowered landings with experimental rocket planes such as the X-1's just after World War II and the X-15's in the 1960s. The venerable Space Shuttle, operated from the early 1980s until the first decade of the twenty-first century, glided back from orbit too, while as early as 1988 the Russian Buran shuttle made a completely automated, uncrewed and unpowered landing on its first (and only) flight.

Today's technology makes it possible for spaceplanes to make a perfect return from orbit on autopilot only, whether it is night or day and regardless of weather conditions.

The spaceplanes operated by Orbital Destinations are all subject to the rules of the Federal Aviation Administration (FAA), which certifies new passenger launch vehicles and licences companies operating these spacecraft. The FAA regulations for spaceplanes are an extension of those for airplanes and have been established to ensure the reliability of the vehicles and operators and the safety of the passengers. The FAA is a government organization, to prevent commercial interests from becoming more important than safety standards. They also supervise the Air and Space Traffic Control that nowadays handles both normal airplanes and spaceplanes during atmospheric flight.

From the security point of view, hijacking a spaceplane or placing a bomb on board is almost impossible as you cannot simply buy a ticket and check in minutes before take off, as is usual for atmospheric airliners. You have to undergo a medical check and several registration formalities months in advance. If you were on any international police or airline's blacklist you would be discovered long before you even started training and you would never get on board a spacecraft. Luggage volume and mass are restricted as well as severely scrutinized to avoid any fire, explosion, or contamination hazards. Taking a bomb or weapon with you into orbit is practically impossible.

Although he has just demonstrated that a flight with a spaceplane is quite safe, the instructor reminds you that at any time during training you can decide not to make the trip, after which you will even get a 50 percent refund of the money you spent on the ticket. Of course, although they don't explicitly tell you, the people of Orbital Destinations prefer to lose some money on a dropout than have someone panicking on board and ruining the experience for the other passengers. In any case, there are always people available, who have already gone through the training, and are waiting for a cheap, last-minute flight.

You have understood and accepted the risks before signing up and don't give it any further thought. You know that the threat of anything going seriously wrong has decreased dramatically since the days that astronauts were launched in delicate capsules and shuttles on fragile rockets, sometimes even with solid propellant boosters that could not be turned off until they burned out!

In the past, astronauts had an estimated 1 in 250 chance of not surviving a Space Shuttle mission. Although at the time this was "par for the course", a similar reliability figure today would mean one unacceptable Orbital Destinations spaceplane crash each year and hundreds of airplane disasters every day! Commercial airline passengers have actually only a 1 in 2 million probability of perishing in an airplane accident. Therefore, every effort is made in both the design and the operation of the spaceplanes to minimize risks and also the insurance costs.

SPACEFLIGHT THEORY

The afternoon is spent on classes that explain some of the principles of spaceflight and the specifics of the spaceplane in which you will be flying, to make you understand what will happen during all phases of your trip.

Basically, your spaceplane is three vehicles combined into one.

During launch it is a rocket, depending on the thrust of its powerful engines to gain sufficient altitude and to accelerate to the orbital velocity of 7.5 kilometers per second (4.7 miles per second). If planes could fly at this velocity, you could go from New York to Los Angeles in 12 minutes, or from London to Paris in less than 1 minute!

In space, the spaceplane is a satellite, kept in orbit by the balance between the Earth's gravity and the centrifugal force caused by the vehicle's circular movement around the planet. In orbit the spaceplane requires only small thrusters for attitude control, as there is hardly any air to decelerate the vehicle. To come back to Earth, the bigger engines are fired opposite to the flight direction, slowing the spaceplane down so that it falls back into the atmosphere. Aerodynamic drag will further reduce the spaceplane's speed, turning velocity into heat that is dissipated into the air.

The vehicle changes into an airplane, with its wings creating sufficient lift for a smooth glide back to the Earth's surface.

GEOGRAPHY

The next day your group (official designation "Crew 22F6", since you are the sixth crew in training in the month of February of the year 2022) heads out to the Geography and Earth Observation Instruction Room. This spacious theater is equipped with a device that projects an image of the Earth on the inside of a huge transparent sphere, so that it resembles the planet as you will see it from orbit. Continents, islands, seas and oceans are depicted in great detail and any kind of cloud pattern as well as seasonal changes in weather and vegetation can be simulated. On the night side of the planet you even see the lights of large cities and industrial complexes.

As the image spins slowly before your eyes, the instructor points out some fascinating deserts, coastlines, and mountains you may want to look out for when you are in space. On a separate screen, pictures of volcanoes, thunderstorms, auroras, and cities are shown to familiarize you with the views and to indicate some of the interesting features you can observe. Knowing what to look for, and where, will increase the fun you will have watching the Earth from orbit.

MEDICAL ISSUES

A class on the medical issues involved in spaceflight is intended to prepare you for what is going to happen to your body in space. On Earth your muscles and bones are continuously subjected to the pull of gravity. This force disappears when you are in orbit and can cause problems to the human body, not because it can't handle the change but because it adapts too well!

On Earth, the heart has to work hard to pump blood up and prevent it from collecting in your legs by the pull of gravity. In microgravity this is no longer necessary but your heart will still continue to pump up blood to your head, resulting in thin "chicken legs" and a puffy face in combination with congestion of the sinuses that makes you feel as if you have a cold. Your body detects that there is an excess of blood in its upper part but considers it as a signal that there is too much water, which has to be excreted in the form of urine. As a result, the amount of blood plasma drops by about 20 percent and the red blood cell count falls similarly. While living in space this will cause no problem, but when an astronaut returns fluids will gather into his legs again, causing a lack of blood in the upper part of the body. A returning astronaut usually feels dizzy for some time and may even pass out owing to a lack of blood to the brain.

For longer flights, there is also the problem of muscle atrophy caused by the fact that you don't have to use much force while floating around. Again, this is not a problem as long as you stay in orbit, but when an astronaut returns he may no longer be able to stand if he didn't exercise regularly during the flight. Professional astronauts staying in orbit for a week or more have to train their muscles for at least two hours a day on a treadmill or similar apparatus. The more they exercise in space, the less time it will take to recover after their return to Earth.

Without gravity to fight, the heart has to do far less pumping work in space and your heartbeat slows down. The amount of heart tissue begins to shrink, since the body no longer needs to maintain the powerful heart muscles required on Earth. Heavy exercise can help here too.

The skeleton is also affected by microgravity. Although bones seem to be solid and inert like your hair, they are in fact alive. Bone material is continuously broken down and recreated, as the body maintains an optimum balance between the bones becoming too heavy (making us slow) or too thin (resulting in a very fragile skeleton). This equilibrium is steered by the forces we put on our skeleton by walking and standing. In space, body parts and especially the legs do not have to carry much weight so the body changes to a balance with thinner bones: bone material is broken down as before, but the production goes into a lower gear. In this way, astronauts can lose up to 1 percent of their bone mass each month. The bone calcium that is no longer needed is dumped with urine, which increases the risk of kidney stone formation as an additional problem.

Studies indicate that this loss of calcium may never be completely restored when the astronaut returns to Earth. Scientists are still looking for effective countermeasures, which may also help patients on Earth who suffer from bone demineralization.

Bones and muscles act as reservoirs for magnesium, which the body needs to fight free radicals (highly oxidizing molecules that can split apart DNA) and to prevent calcium from going inside and damaging cells. The loss of bone and muscle thus also results in a lack of magnesium, which on long flights can be very dangerous.

A way to limit both bone loss and muscle atrophy is to stand on a vibrating plate for at least 20 minutes a day. The plate vibrates at a frequency of about 90 times per second, putting the equivalent of a third of the Earth's gravity on the body. While attached to the device with straps, the hands are free to do other activities, such as conducting experiments or washing the dishes.

Something that does not really have an adverse effect on your health, but can potentially ruin your spaceflight, is space motion sickness. This is what you were tested for by the doctor during the medical test you took before coming to the training camp.

Space motion sickness feels a lot like the terrestrial sickness you may experience in a moving car or on a boat, but it is not directly linked to it: people who never experience any motion sickness problems on Earth may become sick in space, while those who easily get seasick may have no trouble in orbit. About two out of three astronauts suffer from space motion sickness; and one in seven will even experience severe nausea and vomiting (not very nice in microgravity). The discomfort usually begins soon after arriving in orbit but lasts only a few days. Even those who suffer acutely from the sickness will feel better after two or three days. This is the reason why NASA space walks are normally planned no earlier than the third day of a space mission; a sick astronaut may make dangerous mistakes and vomiting inside a space suit can cause a lot of trouble.

Of course, three days is about as long as your spaceflight will last, so to ensure that you enjoy your entire trip, injections are available that release a drug into your bloodstream to counter the symptoms.

Research has revealed that space motion sickness is related to otoconia, microscopic calcium crystals that form a dense mass deep within the inner ear. Accelerations of the head affect the way these tiny crystals bend some 20,000 protruding nerve cell fibers, called cilia. Under normal gravity, the cilia are bent by the weight of the crystals, which helps the brain to determine which way is up. The system functions as a super-sensitive accelerometer, feeding the brain with a steady stream of signals that indicate motion and direction. In microgravity, however, the crystals aren't weighed down, and the brain interprets this as a signal that you are falling (which is in fact what being in orbit is: you are falling around the Earth). However, your eyes tell that you are not falling, at least not with respect to the walls around you. The immediate result of this conflicting information is disorientation: many astronauts suddenly feel themselves upside-down, or even have difficulty in sensing where their own arms and legs are.

If you were on Earth, the mixture of non-corresponding messages the brain receives in microgravity would usually signal some kind of poisoning. Therefore, it triggers a vomiting reaction intended to get rid of the toxin your brain thinks is in your stomach. As a result, you suffer space motion sickness.

After a few days, the disorientation and illness go away. The theory on this phenomenon is that with exposure to microgravity, astronauts suppress vestibular signals (signals coming from the orientation sensors) and become increasingly dependent on vision to perceive motion and orientation. When this adaptation process is completed, the space motion sickness symptoms disappear.

Research has shown that astronauts who instinctively choose to align with their body's vertical axis have much less trouble with space motion sickness than those who take the virtual scene around them as a reference. To stimulate astronauts to refer to their own bodies rather than to their surroundings, they are trained in special devices such as rooms that swing around a person standing on a stable platform.

Experience helps too; most astronauts suffering from space motion sickness during their first space flight experience less problems during their second space flight.

If you remained in space, the changes to your blood volume, muscles, skeleton and vestibular system would not be a problem. In fact, you would become a "Homo Orbitalis," with a thin skeleton, weak legs and a low blood volume adapted to living in microgravity. The trouble would only start when you returned to Earth: you would be unable to stand, you would pass out owing to a lack of blood in the brain; and you would probably break your delicate bones in the process.

As you will certainly not stay in space forever, becoming highly adapted to living in space is not what you want at all. Fortunately, there is no problem in your case: the tourist flight is going to last less than three days, so you will not experience any lasting medical problems when you return from the trip and you will not need to do any special exercises in orbit. (After all, it's supposed to be a holiday!)

THE THIRD DAY

Day three starts with a visit to the realistic, full-scale mock-up of the spaceplane. The two pilots assigned to fly with your crew have joined the group to explain the ins and outs of their vehicle.

The spaceplane is an impressive machine, some 62 meters (203 feet) long with a width of almost 30 meters (98 feet) from one wingtip to another. It's just 8 meters (26 feet) shorter than a Boeing 747 but with its short delta-wings the wingspan is less than half of that of the antique airliner. The design of the spaceplane is quite different from that of a normal passenger aircraft because it is designed to fly at 30 times the speed of sound, i.e. Mach 30, while the 747 hardly made Mach 0.9. From the outside, it looks a lot like a stretched Orbiter or big fighter plane. The giant rocket engines at the back of the vehicle, four large and four smaller ones, are impressive; you could completely fit yourself into one of them.

On a cut-away diagram you can see that most of this vehicle consists of propellant tanks and only a small part is reserved for the passengers and pilots.

The passenger module within the spaceplane is a cylinder about 12 meters (39 feet) long and 5 meters (16 feet) wide, allowing space for 32 tourists. With a small group you head up the stairway built alongside the vehicle to reach this part. There are two decks, each with two rows of eight chairs and windows. There are ladders on the floors, needed to reach all the seats when the vehicle is standing in a vertical position on the launch pad (when the floors are the walls, and vice versa).

The quartz-coated polycarbonate windows are a bit bigger than those you are used to on normal airliners, since of course the view is one of the main attractions of the trip. You will be provided with small binoculars to enhance the experience.

The chairs, with their safety belts, look as if they have been ripped from Formula-1 racecars – i.e. quite sturdy but comfortable. They can be rotated and put in a horizontal position for sleeping. The seats are excellent beds, as in microgravity you will be floating above them, only held in place by the safety belts that double as sleeping restraints.

You notice that in front of each seat there is a flat computer display. This allows you to call up maps and charts to follow the progress of the flight and to help you to identify the parts of the Earth you see beneath. Another nice feature is that you can select your favorite music from a huge database of songs and albums. You can play Europe's "Final Countdown" just before lift-off, the *Star Wars* movie's main theme during launch and Frank Sinatra's "Fly Me To The Moon" when you are lazily staring down at the Earth from orbit. The first-ever space tourist, Dennis Tito, was enjoying himself tremendously, listening to opera music on his portable CD-player when he was on board the International Space Station.

There is also a plug for your digital camera, through which you can connect to the spacecraft's central data memory for storing your recordings. In this way, there is no limit to the number of pictures or the length of video footage you can take. After the flight you can access and download your photos through the Internet.

At the back of the lower level of passenger seats is the personal hygiene room, which includes a 0-*g* toilet and a special enclosed sink that prevents water from spilling out. Food storage and preparation equipment are located in the back of the upper level. The galley includes a microwave oven and a hot water dispenser for "cooking" dehydrated food.

A hole in the ceiling of the first cabin level allows access to the upper deck of the module, where, in the forward section, the two-seat cockpit for the pilot and copilot is located. It has some windows and a couple of video and computer displays, keyboards, buttons and flight control sticks. With respect to the cockpits of airliners and Space Shuttles of the twentieth century it looks very neat and rather empty here. During ascent and landing, the pilots are immersed in a three-dimensional holographic simulation through their headsets and visors, and via this virtual reality system the onboard computer shows them the correct flight path – like a highway in the sky – that the spaceplane needs to follow. Advanced radar and satellite positioning systems work together to create a clear view all around, irrespective of weather conditions. Numerical information and warnings pop up before the eyes of the pilots as necessary, without them having to look at a specific display in the cockpit. Even sounds are used to help to manage information: when another plane is in the vicinity the pilots can hear where it is and whether it is coming closer or flying away. Different sounds are used for different warnings, with the intensity of the alarm increasing with its urgency. There is no need for the many indicators, clocks and displays that used to crowd the cockpits of large aircraft.

Although the technology is now so advanced that spaceplanes are able to make the whole trip to orbit and back without human control, even in emergencies, regulations still require at least one pilot on any type of airplane or spaceplane. Pilots can take over from the automatic systems in case anything unexpected occurs that cannot be handled by the computers. Although they have trained for such a possibility in sophisticated flight simulators on the ground, it is highly unlikely that their piloting skills will ever be required.

In fact, the main tasks of the pilots are not related to flying but consist of maintaining contact with mission control on the ground, acting as tour guides, assisting the passengers in emergency situations, and providing food, drinks and general support to the customers on board. In orbit, there's not much else that the flight crew can do.

The flight crews need not get bored, however, since radiation exposure regulations limit their spaceplane flying careers to only a few trips per year. Shielding to stop the radiation from entering the spaceplane would make the vehicle too heavy to fly, so there is no other solution than to limit the time pilots spend in space.

Exposure limitation regulations, based on a trip of three days in space per flight, imply that the maximum number of flights for spaceplane pilots is only six per year. They are therefore people who usually fly large supersonic passenger planes during their normal working days and are hired as required to command a tourist spaceplane. It is quite popular work among pilots, because not only does it pay very well, but it's also an interesting change of work for a few weeks each year. For space tourists who only stay in orbit for three days and make only a few trips during their lifetime, the total amount of radiation forms no threat to their health.

You will, however, be required to wear a radiation dose meter (dosimeter) during the trip, and those measurements will be included in your personal files to ensure that you are not endangering your health by taking too many spaceflights in a short time. The dosimeter was originally developed for the International Space Station. It contains living yeast cells, bio-engineered to become increasingly fluorescent as they battle to repair the radiation damage to their chromosomes.

A special radiation hazard is caused by solar flares – occasional "storms" on the Sun that expel more radiation than usual for a few hours. Fortunately they can be predicted quite well with the help of Sun observation satellites. A launch can be delayed when a sunstorm is expected, while a spaceplane in orbit has sufficient time to return before the radiation peak arrives.

For the afternoon, your "crew" heads out to the spaceport to witness the launch of one of the spaceplanes. The launch facility is built at the coast to ensure that the vehicles, which are launched in an easterly direction, do not fly over inhabited areas. Although the risk of a serious launch failure is low, the old safety requirements that were established in the last century still apply. Moreover, the ocean provides ample room for emergency water landings, the possibility of which has been taken into account in the design of the spaceplanes.

The spaceport looks much more like an ordinary airport than an old-fashioned launch site: there are large hangars where a spaceplane is currently parked for maintenance, propellant production plants and storage tanks some kilometers away from the launch site, a passenger terminal, a runway for landing, two launch pads, and a control center with radio dishes for communications with the spaceplanes in orbit.

The spaceplanes are also assembled inside the large hangars, from elements transported by train from the manufacturer. Unlike airplanes, spaceplanes cannot take-off from the factory site since the noise and risk levels are deemed to be too high for the populated area in which the plant is located. The fact that the spacecraft can be broken down into manageable elements also facilitates refurbishment.

For propellants, the spaceplanes use liquid hydrogen and liquid oxygen, which are mixed inside the rocket engines to create the pressure that pushes the vehicle up into orbit. Not only is this combination the most powerful available for a conventional rocket, it is also very environmentally friendly since the only exhaust product is pure, clean water.

The four large main engines and an equal number of smaller sustainer thrusters of the spaceplane require some 9,500 liters of propellant per second during the first phase of the flight, when all eight engines are operating. For the whole trip, the spaceplane needs almost 800,000 kilograms (1,760,000 pounds) of oxygen and hydrogen, which is 90 percent of its total launch mass of 900,000 kilograms (1,980,000 pounds). As it would take some 40 tanker trucks or railway tanks to supply the propellants for each flight, it is best to have the propellant plants close to the launch site. Furthermore, the liquid propellants have to be stored at extremely low temperatures, and are therefore much easier kept inside the large, insulated storage tanks near the production plants. Liquid oxygen is made from the atmosphere by liquefying air; liquid hydrogen is produced by reforming methyl alcohol that is brought to the plant by railroad.

What still resembles the old-fashioned, traditional launch centers are the actual launch platforms with their metal skeleton structures, pipelines, water towers, and concrete flame trenches. The sleek metal giant that you see towering into the bright sunlight on Platform 1 is the *Galactica*, one of the four spaceplanes in the operator's fleet, named after spaceships from famous science fiction movies and TV series (the other three are named *Jupiter, Enterprise* and *Millennium Falcon*). The empty vehicle was transported to the launch pad yesterday and there put on a giant rotating mount that lifted it to the vertical launch position. Only then were the liquid oxygen and hydrogen fuels loaded, while the passengers boarded the spaceplane just one hour ago.

The spacecraft will land horizontally like an airplane, but the launch is still vertical, like a conventional rocket. This is to save weight on the undercarriage; for the landing, when the propellant tanks are empty, a relatively small and light landing gear is sufficient, but for take-off the spaceplane is 10 times heavier and would require a very robust and heavy launch gear. Launching vertically saves a lot of weight that can be used to increase the number of passengers.

At the sides of the gantry tower you see a series of cables that run from the level of the passenger cabin to the ground. These are emergency slidewires with baskets attached for escaping as quickly as possible from the spaceplane. The passengers can climb into these baskets in the unlikely event of a dangerous situation just before launch. The slidewires end near a heavy concrete bunker that can shelter all passengers and is designed to withstand the explosion of the nearby spaceplane.

Even from your viewpoint 5 kilometers (3 miles) from the pad, the vehicle seems to be alive – an enormous metal beast fuming vaporized propellant as it is anxiously waiting to be unleashed. A giant digital countdown chronometer near the spectators' stand clicks the seconds away....

5... 4... 3... 2... 1... and liftoff! The simultaneously ignited primary and secondary engines lift the metal giant from the surface. Water is sprayed on the launch pad to dampen the acoustic shock waves that could damage the spaceplane. The water turns to steam that mixes with the rocket's exhaust, creating giant clouds that billow out of the flame trenches and obscure the launch area. Seconds later the vehicle emerges from the vapor clouds, riding on blazing spikes of blue-ish, transparent flames.

Until now the spectacle has been like a silent movie, as it takes 15 seconds for the sound of the launch to reach your ears. First you hear the growling of a rolling thunder approaching; then a snapping, crackling noise, like that of a frying pan, is added. The booming sound diminishes in volume as the spaceplane goes higher and higher. You tilt your head to watch the vehicle climbing, shielding your eyes from the Sun. If there had been any clouds, the spaceplane would have broken through them within seconds, but it is a bright day that allows the spectators to follow the vehicle for a quite a long time.

You see the majestic rocket shrink from a clear winged shape to only a dot of light. After one minute the vehicle has already attained an altitude of 13 kilometers (8 miles) and in six minutes it will be in space, 100 kilometers (62 miles) above the surface of the Earth. The secondary rocket engines will then be fired for some short bursts to increase the orbital altitude to about 250 kilometers (155 miles), where the aerodynamic drag of the atmosphere is virtually zero and the spaceplane can rendezvous with the Orbital Destinations hotel that is permanently in orbit there. One and a half minute after liftoff you have finally lost the small speck of light of the vehicle's rocket engines. A long trail of watery exhaust remains, warped by the wind as it stretches from the launch pad up into the air.

Silent with awe, your group heads back to the bus that will return you to the space camp. Attending a launch is an exhilarating experience and you think of how you will soon be taking a flight in just such a spaceplane. What a thrill it will be! However, first two more days of training await.

EVA TRAINING

The next morning finds you at the Neutral Buoyancy Facility where your crew will be simulating Extra Vehicular Activity or EVA, otherwise known as spacewalking. During your spaceflight you will not actually go outside the spaceplane, since that requires a lot of preparation and is deemed to be too dangerous for normal tourists. Therefore, the EVA simulation is an optional part of the training. It's fun and an interesting complement to the rest of the activities, so you gave up a free morning and put your name on the list.

Apart from fellow crew members, there are some trainees from other crews and a number of day-visitors who only signed up for this experience. The simulation resembles real astronaut EVA training; you will have to loosen some nuts and bolts on a spacecraft mock-up to receive your spacewalk trainee diploma.

The facility consists mainly of a large, 6-meter (20-foot) deep swimming pool with the mock-up resting on the bottom. At the sides of the basin are a couple of spacesuits with a "backdoor" for easy access; they are resting in metal frames that hold them up. Before you enter a suit, your weight has to be measured. Taking your weight and that of the spacesuit into account, the instructors calculate how many pieces of lead they need to attach to you to make your buoyancy neutral. The metal weights are distributed over the suit so that in the water neither your body nor your arms nor legs will tend to sink or float. Apart from the resistance of the water when you move, the effect of hovering in the pool is similar to being weightless in orbit (see Figure 16).

You enter the suit through the back, while your two diving assistants put on their scuba gear. They will be following you, making sure that everything is all right and bringing you to the surface if you have a problem.

The length of the suit's arms and legs are adjusted to fit you comfortably. Next, the pieces of lead and rubber pipes for air and water supply are attached. The air is for breathing, the water is tunneled through tiny rubber tubes inside the suit to prevent you from overheating during the exercise.

The back door is closed and you are attached to a crane that slowly hoists you out of the metal frame and puts you into the pool. The first move you make in the water causes you to spin upside down. Lesson one: every reaction creates an

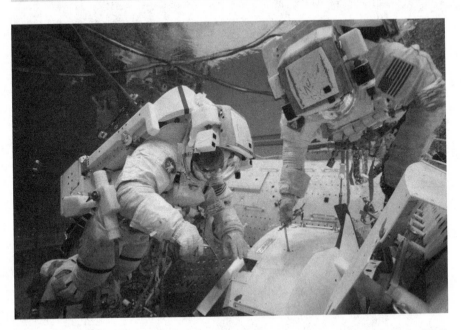

FIGURE 16   Hovering under water in a large swimming pool on Earth, astronauts train for difficult spacewalks. Space tourist trainees could use the same principle to get a feel for microgravity. [Photo: ESA]

equal and opposite reaction, as first described by Newton, meaning that when you move an arm to the left, the rest of your body will move to the right. That's the tricky part of spacewalking, especially when you have nothing to hold on to. Fortunately, the divers are there to take your arms and bring you down to the spacecraft mock-up, or this could have been a very long exercise! At least in the pool you can use the resistance of the water to swim a little. In outer space there is not even air, so there is no way to move in any direction. The only thing you could do there is rotate by swinging your arms and legs.

Spacewalking astronauts are attached to their spacecraft by security cables to prevent them floating away. In case those get loose, they have an emergency return system with small rocket thrusters built into the backpack of their suits.

With the help of your scuba friends you reach for a metal rung on the spacecraft's outer hull and grab it firmly with your thickly gloved left hand. With your right hand you take a cable from your suit and attach it to a ring on the structure. At least now you can no longer drift away.

Your next task is to detach an experiment box with a special zero-torque wrench that does not make you spin when you turn it to loosen the nuts. With some trouble, because of the limited view through the helmet's visor and the thick gloves, you take the tool from the utility belt and go to work. Since you constantly have to balance yourself and because of the stiffness of the suit, the seemingly simple task is quite exhausting. However, you manage to remove the experiment box, which you proudly present to your fellow trainees as you are lifted out of the water.

CENTRIFUGE AND WATER LANDING SURVIVAL

This morning you are going to find out what it feels like to weigh more than three times your normal weight and how to escape from the spaceplane after an emergency landing in the ocean.

A fast centrifuge is used to simulate the accelerations, called *g*-forces, experienced during launch. The normal gravitational attraction of the Earth gives you your normal weight, which is called 1 *g*. The acceleration of the spaceplane during launch will add a further 2 *g*, so you will be pushed into your chair with a force of 3 *g*. The feeling is similar to what you experience in a fast accelerating car but the *g*-forces during launch will be a lot higher.

Because of the relatively limited number of *g*s you will be subjected to and the large number of people that have to be trained each day, the space camp's

Centrifuge Facility is not one of the conventional types with a small cabin at the end of a long rotating arm. Instead a large, round, spinning room is used, with the trainees seated along the circular wall facing the inside.

You and 17 other members of your crew step through the two doors into the 10-meter (33-foot) diameter cylinder and take your seats. Once the instructor has checked that each of you has buckled up correctly, he leaves the room and starts the electric motor at the axis. Slowly you start to spin around and begin to feel heavier and heavier, till the room completes a full rotation once every 2.5 seconds and you are pushed against the wall with a force three times your own weight. It is not very hard to endure, because the seats are quite comfortable and divide the weight evenly over your entire body.

The instructor, watching you from his standpoint above the roofless centrifuge, asks you to lift an arm. That proves to be quite difficut! Your arm feels as if large chunks of lead have been attached to it and you can imagine how hard it must be for pilots and astronauts to push buttons when they are pulling even higher acceleration forces; in the old Gemini–Titan launches of the 1960s, astronauts had to endure up to 8 g! But for the Apollo astronauts flying to the Moon, the maximum launch acceleration was about 4.8 g and for the Space Shuttle astronauts it was similar to what you experience here.

After the centrifuge has slowly stopped and you all carefully get out of the chairs, everyone agrees that the 3-g acceleration of the upcoming trip should not cause any problems.

The water landing survival training is scheduled for the afternoon. As it takes place just one day before your launch, the procedures should be fresh in your mind when you take off.

The training facility consists of a large pool with a mock-up of the passenger cabin of the spaceplane standing in the middle, above the water. No stewards demonstrating life jackets here, but realistic evacuation drills. An emergency landing in the sea will be simulated and your crew will train to get out of the spaceplane as quickly as possible; the vehicle is not designed to float for any length of time and may break up after a hard landing in the water (see Figure 17).

You are wearing your complete flight suit for the occasion, including the integrated inflatable life jacket. Your orange overall resembles those worn by professional astronauts. It has many pockets on the chest, arms and legs, and is lined with Velcro for easy opening and closing. For safety and to keep you warm in case of an emergency water landing, it is made of an insulated, fireproof material. The clothing is quite fashionable, with a definite "spacy" look. It is much better than the gear the first space tourists to the International Space

FIGURE 17 *Astronauts train to use their inflatable lifesuit and survive an unplanned water landing. [Photo: ESA]*

Station had to wear; you remember that Mark Shuttleworth complained about having to fly in old-fashioned Russian overalls that had not changed in design and color since the 1960s.

Some standard equipment you are carrying in your pockets includes emergency rations, a flask with water, a small first-aid kit, a Swiss army knife, a flashlight, and a whistle to signal for help and to find fellow crew members in foggy weather.

Under the outer suit you are wearing a lighter overall, which will be sufficient clothing in the comfy warm environment of the spaceplane in orbit. The light suit fits comfortably, but is tight enough to prevent it from getting caught behind handles and switches. It also has many pockets with Velcro strips that are very handy in microgravity, because anything not in the pockets or otherwise attached will tend to float away.

During launch and the descent to Earth, you will be wearing military boots. They protect your feet and ankles in emergency situations, when you may have to run and jump from considerable heights. In orbit you will exchange them for thick socks, since there you will not walk, but float. This soft footwear also diminishes the risks of hurting anyone when you swing your legs around, especially the first day when your movements may be a bit clumsy.

A light flight helmet with built-in intercom completes your gear and is intended to protect your head during hard landings.

After detailed instructions your crew take their seats in the passenger cabin. First the emergency oxygen system is demonstrated. In case there is not enough air to breathe or if there is smoke in the cabin, you can close the visor of your helmet to cover your face. This will automatically open the oxygen valve, enabling you to breathe normally.

Different from the situation in normal airplanes, the helmets are connected to individual oxygen bottles that can be detached, so you can take the small container with you as you climb to the emergency exits.

The system employs a chemical cartridge to generate oxygen for the user, which results in a much smaller and lighter system than the use of compressed air.

After the oxygen system demonstration, your group waits for the alarm signal to exit through the main door and additional emergency hatches. A siren sounds and you quickly detach the safety belts holding you in the chair. As you are closest to one of the emergency doors, it is your job to open the hatch. The system is simple: all you have to do is pull a thick red cord next to the exit and the door shoots out with a loud bang. The pressurized ejection system throws the door a few meters away, while a large rubber slide is inflated from the hatch down to the water surface.

According to a predetermined order, you and the other trainees egress one by one through the doors and slide into the water. You pull a cord on the yellow life vest to inflate it and with ropes from your suit you attach yourself to fellow crew members to ensure that you do not lose each other in the water. You have all managed to get out successfully but the instructors are not satisfied; it took too long! Your crew goes through the procedure five more times. The last time the cabin's windows are blinded and the lights put out while carbon dioxide smoke, the same stuff pumped out on the dance floors in discotheques to enhance the effect of colored lights and lasers, fills the cabin. This time you pull down the helmet's visor and detach the oxygen bottle to take it with you. You follow the illuminated arrows to find the exits, which fortunately is not too difficult after the previous escapes (see Figure 18).

The water landing survival training completes your crew's preflight preparations. In the evening there is a ceremony where each of you is presented with an Astronaut Trainee diploma and a flight badge with your name and crew designation. The round patch depicts the silhouette of a spaceplane in green with the blue-white disc of the Earth as background. Proudly you fix the Velcro emblem on your flight suit and shake hands with the rest of Crew 22F6. You are all ready to "kick the tires and light the fires"!

FIGURE 18   Cosmonauts practice getting out of their Soyuz capsule, in case it accidentally lands in the water instead of the dry Kazakh steppe. [Photo: ESA]

4

ASTRONAUTS AT SCHOOL

I F you think being an astronaut is mostly about doing heroic, adventurous things in space, you are wrong. Astronauts spend much more time in classrooms and training facilities than actually flying in space. Years of training for a one-week flight is normal. Not only do astronauts have to learn to live in microgravity, but they also have a job to do that involves complicated procedures and technologies. For a successful mission a crew needs to know the spacecraft, the experiments, the procedures, and each other intimately.

From launch until landing, astronauts and cosmonauts have a heavy task list. They do scientific experiments such as growing crystals and plants or measuring the changes microgravity causes in the human body (their own or those of their colleagues). They take pictures of the Earth, operate external robotic arms to attach modules to the Space Station, launch satellites from the payload bay of the Space Shuttle and install new hardware on spacecraft during spacewalks. The Hubble Space Telescope would not have given us such beautiful pictures of the universe if astronauts hadn't fitted a pair of glasses to compensate for an error in the shape of the telescope's mirrors.

But overall, only about a third of the time in orbit is spent on the important activities the astronauts were sent up for. The rest is used for sleeping, eating, personal hygiene and last, but not least, maintenance of the spacecraft. Vacuum cleaning, changing of filters in the life support system, removing condensation and other domestic activities consume about 15 percent of the crew's time.

In microgravity, dust, drops of moisture, food, hairs, and tiny flakes of skin float around in the air. This stuff can start to irritate eyes and throats or enter sensitive equipment, causing malfunctions. What is not removed from the air by filters has to be taken away by the astronauts. In the last years of the operational lifetime of the geriatric Mir space station, the crew was constantly repairing equipment and removing huge blobs of fluid leaking from the cooling system.

On a space station much time is also spent in putting things in order, since there is usually so much equipment on board that without systematically putting things back where they belong equipment can be lost forever. A pen left to float around on its own for only a few seconds may be grabbed by the airflow from the life-support system and disappear in a corner, never to be seen again. Maintenance, cleaning, the placement of equipment in the spacecraft, and preparation of food all have to be part of the training, not only the more evident activities required for experiments and operations.

It is clear that, to become a professional astronaut, a good deal of mental and physical training is required to be able to handle all the onboard spacecraft avionics systems, life-support systems, propulsion systems, communications systems, power systems, computers, science experiments, robotic arms, airlocks, cameras, toilets, kitchen apparatus, etc.

Where and how do astronauts train? Russian cosmonauts train in Star City, not far from the launch site of their Soyuz spacecraft. The first man in space, Yuri Gagarin, was trained there and so was every cosmonaut who followed in his footsteps, including Dennis Tito and Mark Shuttleworth. Tito trained for eight months before the Russian Space Agency qualified him to be a passenger on board their Soyuz spacecraft. The training center has neutral buoyancy pools, detailed mock-ups of the Soyuz capsule and Space Station modules, a huge centrifuge and many medical test facilities and classrooms.

Survival training is done elsewhere; cosmonaut trainees have to spend 48 hours alone with their crew in the middle of a dark forest, using a mock-up of their landing capsule as shelter. The crew of the Voschod 2 actually came down in such a forest after overshooting the intended landing site and had to hide in their capsule from a pack of hungry wolves until they were found by the recovery crew.

NASA's astronauts go to several centers for different parts of their training, while European and Japanese astronauts train mainly in the US or in Star City. Since they do not have their own spacecraft they fly to orbit in either the Russian Soyuz or the Space Shuttle. China has its own training center for their yuhangyuans, which the country launches on board its own Shenzhou spacecraft.

Training Program

The European astronaut training program for International Space Station crew members is based on various training phases.

Basic training provides the astronauts with general knowledge of space technology and science, some medical skills and basic skills related to their future tasks. They learn, for instance, how to operate the Space Station systems. They also take parabolic flights in a KC 135, a large passenger plane that is used to give the astronauts some brief feeling of microgravity. The airplane goes into a steep dive, pulls up and uses the attained speed to follow a parabolic trajectory. During the parabola, everything in the plane is weightless for some 20 seconds. At the end of the short phase of weightlessness, the astronauts have to make sure they are on the floor otherwise they would fall heavily. The process can be repeated again and again, although after more than 20 parabolas most people will be turning a pale shade of green. Hence the nickname of the plane: "Vomit Comet."

Upon completion of the basic training, successful candidates are certified as career astronauts. Basic training takes up to one year.

Advanced training is intended to provide astronauts with the knowledge and skills needed to operate the Station experimental facilities, robotic arms, airlocks and EVA suits. In addition, they learn how to interact with the people of mission control on Earth, who keep track of the mission schedule, offer directions when problems arise and take care of the general well-being of the crew. If the astronauts are, for instance, exhausting themselves too much, the ground control personnel can send them to bed (or at least "strongly suggest that they take some rest"). Upon successful completion of advanced training, after one year or so, an astronaut is ready to be assigned to a specific Space Station mission.

Each astronaut receives basic and advanced training once and specific refresher courses as required.

Increment-specific training takes place after an astronaut has been assigned to a mission. It teaches the crew (and backup crew if applicable) the knowledge and skills required for the specific onboard tasks. During this phase, the crew trains together as much as possible so that the astronauts get to know each other and can form an integrated team. The duration of the increment-specific training is approximately one and a half years and consists of several phases.

From 18 to 12 months before launch the astronauts go through job-oriented training, which focuses on the tasks they will have to perform on board the Station. Astronauts learn how to operate the metallurgy furnaces, the gloveboxes for handling hazardous materials and the incubators for biological experiments. They are taught how to draw blood from fellow crew members and how to make ECGs of their own heart rates.

Job-oriented training is followed by partner element training, which usually lasts until three months before the flight. It is a team-oriented training that concentrates on working with multiple crew members, each of whom has a specific job to do. This includes, for instance, practicing in the enormous Neutral Buoyancy Simulator of NASA's Marshall Space Flight Center, a 23-meter (75-foot) wide, 12-meter (40-foot) deep water tank containing some 5 million liters of water. It is large enough to house a complete, full-scale mock-up of the Hubble Space Telescope on which astronauts practice their repair spacewalks.

During the last six months before a flight, the crew participates in multi-segment training, involving the putting together of payload and systems operations for the entire International Space Station. During this multi-segment training the crew works as a team, sometimes together with mission controllers via integrated simulations.

Once the mission has started and the astronauts are in orbit, they may still continue their preparation with so-called onboard training. The astronauts who operate the complex robotic arms of the Shuttle Orbiter or Space Station may rehearse their procedures in space before performing rather difficult work such as taking new modules out of the payload bay of the Shuttle Orbiter and docking them to the Space Station.

What becomes clear from this description is that most astronaut training does not have much to do with living in space but is much more concerned with actual work. Space tourists would not have to learn how to operate the spacecraft's delicate systems or to do complex scientific experiments. This means that a passenger who only wants to enjoy a few

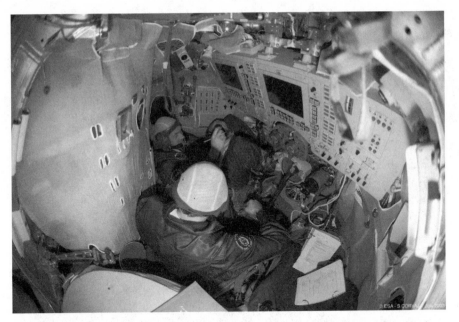

FIGURE 19   *The cramped Soyuz capsule houses lots of systems, screens, dials and switches that a professional cosmonaut needs to know by heart. Space tourists, however, should not be involved with the delicate operation of their spacecraft. [Photo: ESA]*

days of orbital holiday, would not be required to spend years in training. After all, grabbing candies floating around in the cabin or drinking juice through a straw does not take much practice (see Figure 19).

The limited amount of training that would be necessary, if only to prepare the tourists for the flight to ensure that they enjoy it more, should be made part of the holiday. A space camp experience that makes the training fun would actually enhance the attraction; instead of spending a lot of money for a trip of one or two days, tourists could enjoy a complete holiday package of a week or more!

PREPARING THE SPACECRAFT

How do you prepare a spacecraft for flight? It's not just a case of fuel-up and go. Reusable spaceplanes, just like modern airliners, will require services before taking off on a new flight. A regular airplane just needs to be fuelled and have its baggage off-loaded. Maybe there are some empty cans and crumpled newspapers to be removed from the passengers cabin, but that's basically it. That's just what we would like to have for a spaceplane too, with minimal time spent on the ground between subsequent flights. Instead, all current heavy launchers take weeks or even months to be prepared for a launch, involving small armies of ground personnel busy handling the many complex jobs at the launch center.

Ground operations for (partly) reusable spacecraft, like the Shuttle Orbiter, consist of three major phases. The cycle starts with the post–flight operations, which include everything to be done just after the vehicle, or one of its reusable stages, has landed.

First the vehicle or stage has to be "safed": any hazard that the retrieved equipment may pose to the health of the ground crew has to be removed. This is a major concern, since after landing there may still remain some propellant and pressure in the tanks, including toxic liquids and gasses.

When you watch an Orbiter landing, you will notice a whole caravan of cars and trucks following it to the end of the runway. A truck with a huge ventilator on top is positioned behind the Orbiter to blow away toxic exhaust gasses coming from the maneuvering thrusters, while ground crews in protective clothing are busy attaching pipes to the back of the Orbiter to pump away propellant remains. A special "sniffer" is used to determine if there are any toxic gasses left in the air around the vehicle, and only when this equipment has determined that all is safe are the astronauts allowed to disembark (see Figure 20).

FIGURE 20   In contrast to a commercial airliner, a returning Shuttle Orbiter is immediately surrounded by a large ground crew and lots of support equipment. [Photo: NASA]

According to test pilot tradition, the crew will do a walk around their spacecraft for a first inspection. They often notice damaged thermal protection tiles, the replacement of which is one of the main causes for the high price of around $500 million for a Space Shuttle flight.

After safing, the Shuttle Orbiter cannot be towed to the processing facility immediately; the vehicle is literally too hot to handle from the extreme aerodynamic heating during descent. The Orbiter has to stay on the runway for at least four hours to cool off, during which only small, time-sensitive payloads can be unloaded from the crew cabin.

Subsequently, the Shuttle Orbiter is transported to the Orbiter Processing Facility, where the maintenance and refurbishment take place. After a landing at Kennedy Space Center, it can simply be towed there behind a truck. When the weather did not allow a landing in Florida and the Orbiter touched down at Edwards Air Force Base in southern California instead, the whole Orbiter had to be bolted on top of a specially modified Boeing 747 and flown back to the space center. The preferred landing site for any spaceplane should be at the same location as the launch facility to avoid such costly transports, which also delay the preparation of the vehicle for the next launch (see Figure 21).

The Space Shuttle system requires the Solid Rocket Boosters (SRBs) that are ejected during launch to be picked up. Just minutes after liftoff, these 48.5-meter (159-foot) long, white steel tubes land in the ocean with the help of parachutes and are picked up by specially equipped ships, one for each booster. Divers have to go 33 meters (108 feet) under water to put a plug in the nozzle of each SRB, which, just after landing, is floating in a vertical position. Subsequently, air is pumped into the hollow booster, pushing the water out and making the back part come up so that the SRB can be towed behind the ship horizontally. After arrival in the port of Kennedy Space Center the two boosters are washed and disassembled, then put on transport to the manufacturer for refurbishment.

The large brown External Tank that is positioned between the boosters and the Orbiter during launch is not retrieved; it is dropped off just before the spacecraft reaches orbit, and burns up in the atmosphere.

In comparison with all the post-flight operations required for the Shuttle Orbiter, those for a commercial airliner are very simple; they are limited to driving the aircraft to the gate or hangar, powering down the aircraft and off-loading the cargo and passengers.

After the spacecraft arrives at the processing facility, maintenance is required for any part of the vehicle that is reusable. Corrective maintenance of the launch vehicle involves repairs on equipment that is no longer flight worthy ("flight qualified"). For the Shuttle Orbiter, for

FIGURE 21   *After each mission, Shuttle Orbiters are brought to a hanger at Kennedy Space Center for weeks of maintenance work and payload loading. [Photo: NASA]*

example, much effort is spent on replacing damaged thermal protection tiles in the heat shield of the vehicle. Some 35,000 of these have to be inspected one by one, a painstakingly slow and very costly process. As no tile is identical to another, each replacement has to be specially manufactured and only fits in one specific slot on the outer body of the Shuttle Orbiter.

Preventive maintenance consists of repairs that are performed before a part of the vehicle degrades too far to be reliable, although at the time of repair it is not yet broken. The main engines of the Shuttle Orbiter are removed and serviced after each flight – something that should not be necessary for any future spaceplane. Can you imagine bringing your car to the garage every day for servicing?

Spaceplanes of the future should incorporate health-monitoring systems that continuously measure all loads and stresses during the flight as well as the operation of all parts of the vehicle. Upon return of the spacecraft, such a sophisticated integrated sensor system would indicate those parts of the vehicle that did not perform as they were designed to do, or are about to malfunction and therefore require preventive maintenance.

For any machine, major parts such as wheels, structures, wings, and engines become unusable after a certain time. They wear out. Refurbishment will then be necessary, involving disassembly of the vehicle and taking it out of operation for a longer time than during routine maintenance. For aircraft, major overhauls are performed after a certain numbers of flights – for much the same reason that your car is serviced after you have driven a certain number of kilometers (although you have never covered 4.7 million kilometers (2.9 million miles) in a week, as the Shuttle Orbiter does). For the Orbiter, you could argue that major refurbishment is required after each mission, since the maintenance is so extensive. Clearly not the way to go for tourist spaceplanes.

The Shuttle Orbiter alone (apart from the rocket boosters) requires a team of some 90 people working more than 1,030 hours in total on maintenance and refurbishment after each mission, costing about $7.5 million per flight. Not included in this is the work performed on the three main Orbiter engines, which is done separately in the engine shop of the massive Vehicle Assembly Building. Modern rocket engines are not nearly as reliable as airliner jet engines and require much more maintenance, illustrated by the fact that rocket engine operational times are measured in minutes, while jet engines are operational for hundreds of hours without problems.

Another related difficulty is that maintenance requires many systems (such as electronics, power supply, and air-conditioning) to be active during inspection and repairs. Long maintenance periods thus also result in long power-on times for various onboard systems, which means that the equipment is wearing more quickly and the chance of failures increases while the vehicle is not even in use. In the case of the Shuttle Orbiter, the power-on time on the ground is much longer than the power-on time during actual flight: 1,450 hours against 200 hours! Many systems in the Orbiter are therefore basically aging while they are only standing on the ground.

It is clear that with the type of maintenance and refurbishment scheme that is applied to the Space Shuttle, such spacecraft will never be able to carry tourists into space at acceptable prices. One of the main technical challenges for future reusable launchers is to dramatically lower the amount of time and money spent on maintenance.

After all those repairs, is the spacecraft ready to go into space? The answer is "No." It may be in good shape again, but still has to be prepared for the next launch during the so-called preflight phase. Preflight operations begin with the flight segment mating, which means putting together the various components of the Space Shuttle (for instance, the Orbiter, the External Tank, and the two Solid Rocket Boosters). Next the payload, such as a satellite or experiment module, is installed in the vehicle's payload bay (see Figures 22 and 23). Then the vehicle may need to be transported to the launch pad, where propellant loading, tank pressurization and the activation of onboard avionics, power systems, and communication equipment takes place (see Figures 24 and 25).

Also, the ground segment needs to be prepared: the launch team gets ready in the control center and tracking facilities. Emergency services at the launch center and the abort landing sites are put on standby. The Shuttle Orbiter has, for instance, a number of emergency landing runways in Europe and Africa where NASA crash and rescue teams stand ready during launch, hoping that they will never be called upon to act.

Before launch the zone around the launch pad, as well as the part of airspace through which the vehicle will travel, has to be cleared of all personnel who have no reason to be there. Rocket launches are sometimes delayed because people fly over the launch center in sports planes, or because fishermen accidentally take their boats into the safety parameter. Falling stages or debris from an explosion could hit people who are too close to the launch pad or under the trajectory of the rocket. For this reason, rockets from Japan's Tanagashima space center can only be launched for 130 days each year, the rest of the time the area is reserved for the fishing fleet operating in the sea near the launch center.

Astronauts enter their spacecraft at least an hour before launch, but on many occasions there are launch delays that force the crew members to wait in their seats for hours. America's first astronaut, Alan Shepard, was also the first to have to endure a launch delay. In fact, he had to wait so long that he eventually needed to go to the toilet. As this would have involved unbolting the capsule door and putting the launch preparations on hold – causing an even longer delay – it was decided that the unfortunate astronaut should relieve himself in his suit (which was never intended for such abuse). Shuttle astronauts now wear diapers for such occasions. Russian cosmonauts do not need them, because they are simply launched on time.

An important issue for maintenance and preflight operations is whether the vehicle can be processed horizontally or has to be prepared vertically, like many current expendable launchers. Horizontal processing is easier and less labor-intensive, because it facilitates access to all parts of the vehicle. This is an important factor when the flight rate is high and the turnaround time (processing time between flights) needs to be low. The

FIGURE 22 *The huge Vehicle Assembly Building, originally built to house the Saturn V moonrockets, is now used for the vertical integrating of the Shuttle Orbiter, External Tank and the two Solid Propellant Boosters. [Photo: H. Joumier]*

FIGURE 23   The large first stage of the European Ariane 5 launcher is standing upright in the assembly building at the launch site in French Guyana. The two solid propellant boosters, the upper stage and the satellite payload have still to be added. [Photo: M.O. van Pelt]

FIGURE 24   In a vertical position, the complete Ariane 5 launcher is slowly driven to the launch pad on this huge rail transport table. [Photo: M.O. van Pelt]

FIGURE 25   *An Ariane 5 launcher standing on the launch pad, surrounded by a complex of water towers, lightning poles, exhausts ducts and support buildings. [Photo: M.O. van Pelt]*

Russians have always favored horizontal processing of their rockets, even the very large ones like the giant Energia.

Another key factor is the type of propellant used. Kerosene can be handled as at any airport, but cryogenic propellants with extremely low temperatures such as liquid oxygen and liquid hydrogen (cold enough to freeze a piece of rubber instantly, making it as brittle as glass) require on-site production, intensive cooling, insulation of piping, vapor removal, and complex handling procedures. Toxic propellants like hydrazine require rigorous safety measures and special equipment to protect the health of the people working on the launcher.

In comparison, the preflight operations for a commercial airliner are much simpler than for space launch vehicles. They mainly involve kerosene fuelling, cargo and baggage loading, and passenger boarding. The flight crew sit in the cockpit just before flight, power up the plane, wait for clearance from the traffic control tower and the airliner is ready for take-off within minutes. Such efficient procedures make it possible for Miami airport, for instance, to handle 90,000 passengers per day.

This sets an example for what our goal should be for the operation of future spaceplanes, especially those to be used to carry large numbers of passengers into orbit every day.

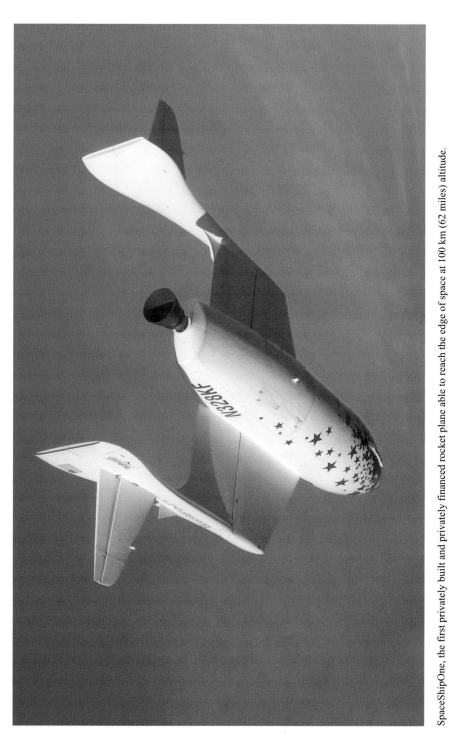

SpaceShipOne, the first privately built and privately financed rocket plane able to reach the edge of space at 100 km (62 miles) altitude.

The Russian Soyuz launcher is driven horizontally to the launch pad by rail. Once it gets there, it is put upright and filled with propellant just before launch. (Photo: ESA)

A Space Shuttle stands ready for launch. Part of the launch tower can be rotated to cover the Orbiter's back for final work on the satellite, space station module or other cargo in its Payload Bay. (Photo: NASA)

SpaceShipOne hanging under its White Knight mother ship, from which it is dropped at high altitude. (Photo: Scaled Composites)

The special suits that Shuttle astronauts wear can help them to survive fires, cabin depressurization and water landings, but cannot save anyone in case of explosions during launch or the problems during reentry. (Photo: NASA)

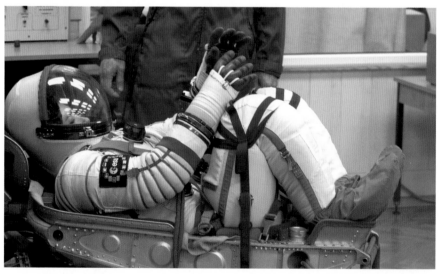

Launches and landings with the Soyuz system can be tough; cosmonauts need to stay on their back in chairs specially fitted to their bodies to help cope with the vertical accelerations and shocks. Future specialized vehicles for space tourists should offer a milder flight experience and use standardized chairs for everyone, like those used in the Space Shuttle. (Photo: ESA)

A Soyuz launcher lifts off from the Baikonur cosmodrome. First used in 1967, the Soyuz launcher and capsule still form the backbone of Russia's human spaceflight program. (Photo: ESA)

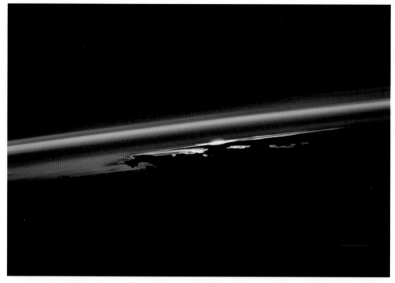

Sunsets as seen from orbit are truly spectacular and can be seen sixteen times per day. (Photo: NASA)

From orbit the curvature of the Earth can be clearly seen. It is also apparent how thin and fragile the atmosphere really is. (Photo: NASA)

The possibility to see one's own country from space will be a major space tourist attraction. Here The Netherlands can be seen, as photographed by Dutch astronaut André Kuipers. (Photo: ESA)

Eating in space will probably never be fine dining, but in microgravity it can be a lot of fun. (Photo: ESA)

A Shuttle Orbiter lands at the Kennedy Space Center, where it was also launched. A small parachute helps to brake the spacecraft on the runway. Airplane-like landings can be made routine and are therefore the way to go for future space-tourism vehicles. (Photo: NASA)

A large, wheel-shaped space hotel could be constructed out of Space Shuttle External Tanks. Rotation would create artificial gravity to make living inside more normal. (Photo: Space Island Group)

5

LAUNCH

YOU are seated at a long table in the restaurant of the Vostok Hotel, enjoying a traditional steak-and-eggs breakfast on the day of your launch. Although this is not really your idea of a light breakfast, you decide to stick to the old tradition and happily chew on.

Unlike this morning's menu, the other meals of the last few days have been specially put together to keep you healthy in space. They contained mainly low-fiber food to minimize the need for the somewhat awkward space toilet. Even though you now know how it works, the use of it takes a lot of time and, for weight reasons, there is only one on board the spaceplane.

The entire crew is present for breakfast, no latecomers today. Everyone is wearing a bright orange flight suit, with the mission's artistic patch proudly worn on the chest. The people who were a little nervous before starting training have become much more confident in the safety of the spaceplane and in their own abilities to handle the flight and life in space.

The mood at the table is one of excited expectation. Through the training you have gotten to know the crew members well, made friends and enjoyed others' advice. You will be going into orbit with a happy, tightly knit group. This is quite important, as you will be locked up in a relatively small space with these people for three days.

The pilot and copilot have joined too, each sitting at an end of the table to answer any questions you may still have. They appear cheerful, yet are professional and serious about the upcoming flight. This will not be their first mission, but they still look forward to it. Someone asks if spaceflight ever gets boring, upon which the copilot answers that he never met anyone who got tired of watching the Earth from orbit. He says that going into space for the second time was even more thrilling than his first flight, because the second time he was aware of the marvelous views that were awaiting him.

The pilot says that most of the year he flies one of those new, Airbus Mach Two passenger planes. It is rather exciting to fly these machines, but every time he is cruising high above the clouds he has to resist the longing to pull up the nose of his plane and fly up into space. The big plane usually does not fly higher than 20 kilometers (12 miles), where the Earth still looks rather flat and close by. It just does not compare to spaceflight.

After breakfast it is time to put your survival equipment, radiation dosimeter, and digital camera in your pockets. You help some less flexible crew members to fasten things in those hard-to-reach places and take your helmet from the rack.

Your limited personal luggage has already been brought on board the spaceplane and stowed in the passenger cabin by the ground crew. You only packed your diary and some small souvenirs, such as mission patches to give as presents to your friends back home. The time you will be spending in space is too precious to fill with anything other than looking out of the window and enjoying microgravity.

You perform some last minute checks on your gear and then you're off.

As you walk out of the door into the bright morning sunlight, you are greeted by friends and family who have grouped in front of the hotel to wish you a good flight. Your crew proudly poses on the stairs, so that everyone can take a picture of the brave-looking group. Although space tourism has been around for quite some time now, it is still special enough to make you feel like an explorer about to venture into new territory.

You wave at friends and family as you board the large aluminum bus to the launch pad. As the bus leaves the space camp and heads for the awaiting spaceplane, your relatives and friends move to the observation site to watch you go into space.

As the bus reaches the spaceplane, you see that the ground crew is already finalizing the operations at the pad. While you are entering the launch area, everyone else is leaving!

The bus stops just a few tens of meters away from the launch gantry, which is partly stained dark red by the frequent exposure to rocket exhaust. Exiting the bus, you look up at the giant winged rocket that will soon hurtle you hundreds of kilometers into space. Up close, the *Jupiter* is even more impressive than from the visitors' stand; a marvel of technology. You see the vapors from the rocket engine pre-cooling escape at the bottom of the spaceplane. You hear metal moan and shriek under the pressure in the tanks and the huge temperature differences between the fluids inside and the already warm Florida air.

After a video and photo session with the launch vehicle in the background, you proceed in groups of eight up the gantry tower in one of the large elevators. Slowly you slide past the white side of the spacecraft that seems to be spray-painted with slight streaks of matt black grime from past reentries. Bright patches of snowy frost silhouette the contours of the cryogenic propellant tanks inside, covering the yellowish white hull of the spaceplane.

The elevator finally stops at the level of the passenger entrance hatch. A narrow bridge attached to the gantry tower brings you to a box-like room that connects to the spaceplane's hatch. Some ground crew technicians wait to help you into the vehicle and get settled inside. They check your suit and assign you a seat at the upper level.

Jokingly, one of the technicians asks you for your flight ticket, which reminds you of an anecdote you heard during training: Gordon Cooper, one of the first astronauts, was presented with a special coupon just before his first flight, inviting him to take the "ride of his life" for only 25 cents, payable at the launch pad. Just before the first launch attempt for his Mercury flight, Cooper presented the leader of the launch pad technicians who had given him the coupon, the 25 cents on a memorial plaque. According to this story, Cooper may be viewed as the very first space tourist!

As the spaceplane is standing in a vertical position, you have to use the ladders attached to what in orbit will become the floors to reach your seat. After you have climbed into your chair, you are in a 90-degree inclined sitting position, lying on your back and looking upward. This is the optimal position for the body to withstand the launch accelerations.

The ground crew help you to fasten the seatbelts and adjust the headpiece of the chair so that it supports your head correctly. Next they plug in the intercom system to enable you to hear messages from the pilots and other crew members. The speech system is automatically activated when you talk. Some crew members are babbling with excitement, or try to ease the tension with jokes, but most are silently awaiting the launch.

The spaceplane looks quite new and pristine inside. Maintenance standards within the Orbital Destinations company are very high, and not only for critical spaceplane systems. Any loose screw could turn into an unguided projectile during accelerations and could get stuck in dangerous places in microgravity. In the early days, there had once been a problem with a Shuttle Orbiter's airlock door that wouldn't open because something small got trapped in the hatch's mechanism.

A neat looking passenger cabin also serves to create the right sense of safety for the tourists; even bad covering on a chair, although not critical, may give the impression that the spaceplane is not well maintained and therefore dangerous to fly.

You press a key on the display in front of you and the flat computer screen lights up. It currently shows you the vehicle's status and the countdown chronometer that is ticking away the minutes and seconds before launch. With 44 minutes still to go, the status of all spacecraft systems shows green.

The pilots were the first to enter the cabin and are now busy going through their elaborate checklist. They ensure that all the switches in the cockpit are in the right position for flight, that all propellant and cabin oxygen tanks are full, and that there is no loose equipment lingering in the cabin. The crew members are asked to check their seats, seatbelts, and emergency life-support systems. One by one you report in through the intercom system to check the proper working of all communications equipment. Once more, the copilot runs you through the safety and emergency procedures as a refresher of what you learned yesterday.

Unlike in the past, the last pre-launch procedures are now performed by the pilots and the onboard computers and no longer by large launch teams at the mission control center. Outside the spaceplane only a small number of people are directly involved in the last minutes of the countdown, mainly as backup for monitoring the spacecraft's systems and to keep an eye on the launch pad, the weather, and the base's airspace above. This is very similar to the duties of people in the control tower at a regular airport.

You plug in your digital camera and snap some shots of the launch pad through the window. There is not much to see at this moment, and nearly all ground crew workers have left the site by now as most actions continue automatically, under the command of the onboard launch control computers.

To ease the tension of waiting for the clock to reach zero, you click on the computer menu to select some of your favorite music. (Even Yuri Gagarin listened to music while waiting for launch.) Soon Led Zeppelin's "Stairway to Heaven" rings through your ears. Just as The Police start "Walking on the Moon", the voice of the pilot overrides the music to inform you that everything is "go" and that the hatch has been closed.

The ground crew workers who have helped to settle everyone in are leaving the launch pad, but the crew-access arm between the gantry tower and the spaceplane remains connected to the hatch until just before launch. This enables you to escape if an emergency occurs during the final phases of the countdown.

As the music continues, you finally start to realize you are really going into space. Until now it has been a bit like a realistic science fiction movie or a virtual reality game, but here you are inside this spaceplane, all buckled in, hatch closed, ready to go.

"T minus 5 and counting", one of the pilots informs you, 5 minutes until launch. In the past, especially with the Space Shuttle, there were many holds in the countdown to fix sudden problems and malfunctions. However, the launch of a spaceplane is now so routine that the clock has not been stopped even once.

T minus 3 minutes and 30 seconds; a message on your screen tells you that the spaceplane is now on internal power and no longer using electricity coming from outside.

T minus 3 minutes; the engines' gimbal check swivels the large rocket nozzles around to ensure they can move freely to steer the spaceplane after launch (during launch the spaceplane is maneuvered this way, because airplane ailerons, rudders, and elevators would be useless outside the thicker part of the atmosphere).

T minus 2 minutes; a green status box on your screen indicates that the liquid oxygen propellant tank just reached its nominal pressure. You turn off the music to experience the full silent intensity of the last seconds of the countdown. It's like waiting for the clock to call midnight on New Year's Eve, but today the fireworks will be bigger and you will be on top of it.

T minus 1 minute, and the liquid hydrogen tank is now pressurized too. T minus 28 seconds; the hydraulics steering the rocket engines are activated and the engines swivel into launch position.

T minus 7 seconds; you feel a low-frequency vibration going through the vehicle as the rocket engines are started and build up thrust. Giant hold-down clamps still keep the spaceplane securely connected to the launch pad until it has been verified that all engines are functioning correctly and have achieved sufficient power. If not, the engines could be stopped safely and the plane would remain on the ground.

T minus 0 ... Lift-off!

The clamps let go of the roaring monster and you feel the acceleration gently building up as the raw power of the mighty engines is released. At first the spaceplane goes straight up to clear the tower, but after 7 seconds the vehicle rotates toward the East to align itself with the optimal launch trajectory. The rotation gives you a good view of the ground through your window. The launch site is getting smaller and smaller surprisingly fast! A look at the computer screen informs you that the velocity of the spaceplane and the *g*-levels are both increasing. Adrenaline is racing through your veins.

A small good-luck charm, a puppet monkey astronaut, is dangling from the chair in front of you. As in the old Russian Soyuz capsules, it will indicate microgravity when it starts to float, but now it is oscillating like a pendulum with higher and higher frequency as the acceleration of the spaceplane increases (see Figures 26 and 27).

FIGURE 26   *Gemini 11 lifted off in 1966 with two astronauts on board. People still go into orbit in very much the same way, and will probably do so for the foreseeable future. [Photo: NASA]*

FIGURE 27   *A Space Shuttle leaves Kennedy Space Center. The pillar of smoke indicates the huge amounts of propellant that are being burned for this in a very short time. [Photo: NASA]*

After 40 seconds you go supersonic, about 6 kilometers (3.7 miles) above the Earth's surface. The rumbling noise of the rocket engines fades away as the spaceplane outruns its own sound waves traveling through the air outside. The vibrations you feel through the seat remain however, and, together with the now constant acceleration, they keep reminding you that you are still going somewhere very quickly.

A minute after take-off and you are going through max-Q, or maximum dynamic pressure: the atmospheric density and the velocity of the spaceplane bundle their forces in an attempt to tear the vehicle apart. "Shaken like a rat in the jaws of a terrier" is how an astronaut once described the experience, but on this spaceplane it's not so bad. Vibrations reach a peak but slowly subside as the air outside gets thinner and thinner. The acceleration picks up again and rises to a maximum of 3 g.

Already higher than you have ever been in any airplane, you start to notice the curvature of the horizon in the distance. From this altitude the Earth begins to reveal itself as a sphere. You no longer feel like you are in a plane flying over a flat landscape, but actually in a spacecraft turning around a planet.

Suddenly the g-level drops to 2: the four large booster engines have been stopped and the spaceplane continues its ascent on the remaining four sustainer thrusters.

You are now 60 kilometers (37 miles) high, a little more than 3 minutes into the flight. With the main engines out and most of the atmosphere left behind, the flight is now very smooth and silent. The sky slowly turns from blue to jet-black as you are entering outer space.

You continue to accelerate to orbital velocity for another three and a half minutes. Then suddenly the weight pressing you into your seat disappears and all vibrations stop. The low, buzzing hum has gone: obviously the sustainer rocket engines have now been stopped too. All that is left is the continuous background noise of the air fans and life-support system valves and pumps.

The little monkey is no longer hanging down, but slowly floating in front of your eyes, its cord slack and coiling up. You are in orbit!

6

THE SKY IS NOT
THE LIMIT

WHEN did people start to think about leaving the Earth behind, to go and explore what's beyond the clouds? There are many ancient tales about people trying to fly to the Moon. Usually the means of transportation in these stories is very implausible, such as a large collection of geese pulling a cart, a ship caught in an extremely heavy upward storm, or a collection of bottles filled with morning dew (a contraption of Cyrano de Bergerac that was supposed to go up with the dew evaporating in the Sun).

The first person ever to try to launch himself into space with the use of rockets was, according to legend, a Chinese Mandarin nobleman named Wan Hu who wanted to visit the Moon goddess some 3,000 years ago. He strapped a series of 47 bamboo rockets to a chair balanced by two kites and disappeared in an enormous explosion, never to be seen again. He used the right technology, but the implementation left much to be desired.

In the nineteenth century, famous writer Jules Verne described a trip to the Moon in a grenade-like capsule launched by a giant cannon in his books *From the Earth to the Moon* and in the sequel *Around the Moon*. The three bold explorers and a couple of animals experience microgravity, fly around the Moon, ignite a series of rockets that bring them back to Earth, and land in the ocean to be recovered by a navy ship.

There is an eerie resemblance to the Apollo missions that were launched to the Moon in the late 1960s and early 1970s: the Apollo spacecraft has a similar shape as Jules Verne's projectile and also had three men on board. The Apollo astronauts came back from the Moon by use of rocket engines, much as described in Verne's story. The Apollo landing was also at sea, where the capsule was recovered by a navy ship, and the enormous cannon of Verne's story was located in Florida, not too far from where the Apollo launch pads would stand one hundred years later.

Verne's astronauts would have been flattened into large pancakes by the sudden and extreme acceleration of the launch out of the monstrous cannon. Verne understood that he had a structural problem and found the solution in the use of a loose wooden cabin floor resting on a thick layer of water. During launch, the acceleration would squeeze the water through holes in the bottom of the grenade-capsule and thereby cushion the extreme g-forces the occupants would experience. In reality, this would not have helped much.

The French silent movie *From the Earth to the Moon* of 1902 was the first film ever to use special effects to show life in space. Inspired by Verne's books, it depicts a group of astronomers who launch themselves to the Moon with a cannon. They actually land on its surface and find the place inhabited by aggressive aliens who attack the bold adventurers. Fortunately, the Moon creatures explode into clouds of smoke when hit with a stick, which enables the explorers to escape. The men barely make it back to their hollow grenade, which subsequently falls from the Moon back to Earth. It lands in the sea where the heroes are rescued by a ship.

The first person to suggest realistic, scientific theories about how humanity could reach for the heavens and even live in space was a deaf Russian schoolteacher by the name of Konstantin Tsiolkovsky (1857–1935). He understood that only rockets would be able to reach the extreme altitudes and velocities required for spaceflight. Rockets, he reasoned, do not require atmospheric oxygen to sustain combustion, since they carry their own oxidizers. Furthermore, rockets do not need air to accelerate through a propeller because they produce their own gasses.

Tsiolkovsky understood the physics of rocketry and formulated the basic mathematical equations describing the workings of a rocket and the dynamics of its flight. In 1903, the same year the Wright brothers' first motorized plane took off, he invented the method of "staging" to increase the performance of launchers. He even sketched rotating space stations that used greenhouses to maintain the onboard atmosphere and thought of an airlock so that astronauts could leave their spaceships.

Tsiolkovsky did not actually build rockets, but his work inspired a number of young Russian rocket experimenters. In the 1930s they began to put Tsiolkovsky's theories into practice by building small experimental rockets that used liquid propellants.

Also in other countries scientists and engineers started to dream about launching people into space. In 1926 the American Robert Goddard was the first to fly a liquid-fueled rocket. Until then all rockets had been based on solid propellant, a less complex but also less efficient technology: solid propellant rockets have a high mass relative to their thrust and, moreover, cannot be throttled or stopped before they burn out completely. Liquid propellants are therefore essential for the control or landing of spacecraft.

In 1935 Goddard's rockets became so sophisticated that they broke the sound barrier and reached 1.5 kilometers (0.9 mile) high, but the US government did not appreciate the possibilities of the new technology. During World War II all they did was assign Goddard a contract for the development of a starting rocket to help carrier aircraft to take off from ships. The lack of interest in rocket development was partly due to Goddard himself, who preferred to work in secret and had no desire to publicize his work.

In Germany, Hermann Oberth better understood the need to tell the world about the new possibilities that rockets could bring. In 1923 he published a book called *The Rocket into Planetary Space*, which inspired a group of young German engineers to start doing their own experiments.

A converted "Ente" sailplane became the first rocket-powered aircraft in 1928, while in the same year car manufacturer Fritz von Opel achieved 200 kilometers per hour (124 miles per hour) in a rocket car and reached 153 kilometers per hour (95 miles per hour) flying a glider fitted with solid propellant rockets.

In 1931, Germany's first liquid propellant rocket left a proving ground near Berlin, while two years later Viennese Professor Eugene Sänger published a book called *Raketenflugtechnik* (rocketflight technology) that contained a concept for a high-speed, rocket-propelled research aircraft.

Soon the new developments in rocketry caught the attention of the German military. The humiliating Treaty of Versailles, which Germany had been forced to sign after its defeat in World War I, prohibited the German army from developing and employing long-range artillery guns. German army leaders sought to circumvent this rule by supporting the young rocket scientists, lead by 20-year-old Wernher von Braun, in developing powerful rockets as an alternative to large cannons.

The dreamy engineers envisioned launchers that could take people into space; the military wanted long-range ballistic missiles. Both parties needed the same rocket technology to achieve their goals, and while the engineers had the brains, the military had the money. A Faustian agreement was signed between the army and the space enthusiasts, in which the army provided the financial and logistical support to enable the engineers to develop large, increasingly sophisticated and powerful rockets.

Until World War II rockets had been small and incapable of carrying much or going very high, but with the support of the Nazi government von Braun's team developed a 14-meter (47-foot) high monster capable of launching a 738-kilogram (1,630-pound) warhead a distance of 418 kilometers (260 miles). In 1942 they launched their first rocket into space, reaching a height of 80 kilometers (50 miles), a distance of 193 kilometers (120 miles), and attaining a velocity of over 5,300 kilometers per hour (3,290 miles per hour).

The Peenemünde group of scientists created the A-4 rocket, which was soon renamed Vengeance-2 by the Nazis. Thousands of V-2 rocket bombs were launched to wreak havoc on the cities of London and liberated Antwerp. The dream of spaceflight had turned into a nightmare.

As the war turned against the Nazis, they even devised plans for the development of a large intercontinental ballistic missile to reach the United States from Germany. Because of the high inaccuracy of the electronic guidance systems of the time, the huge launcher would have had a pilot on board to steer the rocket's upper part to the target. He would first be guided by radio beacons on surfaced German submarines in the Atlantic Ocean and at the last moment would lock the target into an optical sight. The pilot would then bail out just before impact – death or imprisonment being his only other options.

Although this "Project Amerika" was never completed, it is interesting to note that had the developments progressed as planned, Germany would have been able to launch someone into space as early as 1946.

Just after the war the race was on to find the German rocket engineers, blueprints and hardware. Soviet, British, and American military forces searched the smoking ruins of the Nazi empire to get their hands on the secret technology. Although officially the Allies were still friends, the first chills of the coming Cold War could already be felt, and both the Soviets and the West wanted to be ready for the next power struggle.

The United States managed to get the best scientists, because von Braun and most of the leading figures of the Peenemünde team surrendered themselves to the Americans. Furthermore, during a daring and highly secret move the American army managed to smuggle 341 rail carts of rocket equipment out of an underground factory in the Harz mountains, just before the area had to be turned over to the Soviets.

The Americans now had everything they needed: the rockets, the documentation, and most of the experts. Nevertheless, the Russians managed to catch a few German rocket experts and some rocket parts as well, while the British also obtained some hardware.

America and Russia soon moved the German engineers to their own development centers and put them to work on new ballistic missile projects. From an analysis of the German rocket achievements, the former Allies had concluded that interballistic nuclear missiles were possible. With the Cold War just starting, the development of large rockets became a top priority in America and Russia.

The Soviets put a lot of effort into building large rockets capable of launching their rather heavy nuclear bombs to the United States. In 1957, the state of Soviet rocketry was so far advanced that they could use a converted ballistic missile to put a satellite into orbit. On October 4 a small metallic ball called Sputnik was successfully launched. As it circled the planet, it sent out a beeping signal that could be received by radios all over the planet. The attention and admiration of the world was overwhelming; Sputnik was a bigger propaganda success than even the Soviets had ever imagined (see Figure 28).

America was taken by complete surprise. Nightmares of Soviet nuclear missile platforms in space and on the Moon made the US scramble to get its own satellites into orbit. The problem was that the American nuclear bombs were much lighter than those of the Soviets, so the US had only developed relatively small ballistic missiles.

The US Navy's fragile Vanguard rocket, hoped to be America's answer to Sputnik by launching a puny satellite (Soviet leader Khrushchev called it a "grapefruit"), toppled over and exploded on the launch pad in front of the television cameras.

Fortunately, von Braun's team had not lost its vision of launching satellites while working on rockets for the US Army. Unofficially and secretly, the German experts had converted a ballistic missile into a satellite launcher. Now, with the whole nation in a hurry to catch up with the Soviets, von Braun and his men were given the go-ahead to launch their Juno-1 rocket. The flight was a complete success and gave America back some of its confidence in its own technological capabilities.

FIGURE 28 *The Soyuz rocket used today to send cosmonauts and the first space tourists to the International Space Station is based on the launcher that put the first satellite Sputnik into orbit in 1957. [Photo: ESA]*

Meanwhile, the Soviet Union had launched a larger Sputnik 2 with a dog on board! The dog, Laika, could not be retrieved and died in space, but this made it clear that it was only a matter of time before a Soviet would be put in orbit.

In spite of America's efforts to catch up, the USSR once again stole the show when it launched Yuri Gagarin. Shouting *Poyekhali* (Let's go!), Yuri became the first human being ever to experience the thrill of a launch with a large rocket.

He could not relax much, because any second the emergency ejection chair could have been activated if something went wrong with the launcher. He had to tense his muscles constantly because the ejection jolt could otherwise have crunched his spine, snapped his neck or ripped off his legs if his knees snagged behind the hatchway's rim. The g-loads pulled his face muscles and made it difficult to speak through the microphone. When the rocket's four side-slung boosters were dropped, the vehicle paused its accelerations for a few seconds and Gagarin was thrown violently forward against his straps.

After three minutes he was so high above the Earth, nearly out of the atmosphere, that the aerodynamically shaped cap was no longer required and was pyrotechnically ejected. Now the window was no longer blocked by the protective fairing and he was finally able to see something outside. Gagarin caught a glimpse of the sky, which was already dark blue because of the altitude.

After five minutes there was another jolt as the exhausted central stage was dropped, but soon the acceleration picked up again as Gagarin rode the small upper stage into space. After nine minutes, Yuri and humanity had arrived in orbit. Humanity's first cosmonaut reported "Weightlessness has begun. It's not at all unpleasant, and I'm feeling fine."

Gagarin's flight showed that humans could go into space, and that the experience is not particularly stressing; the short period of weightlessness had not affected Gagarin badly in spite of what some doctors had predicted.

Soon America was putting up its own astronauts in the tiny Mercury capsules, and they didn't find weightlessness to be a discomfort. On the contrary, they rather enjoyed observing things float through their capsule! Furthermore, the astronauts had insisted on having a window in their spacecraft and found that the view of Earth from space was absolutely breathtaking.

The cramped Vostok and Mercury capsules of the early 1960s were soon followed by the larger Gemini, Soyuz, and Apollo capsules, and in the 1980s by the much larger Space Shuttle. More volume meant that astronauts could move around more freely, and they soon discovered that weightlessness was fun and easy to get used to. Longer flights meant that more sophisticated life-support systems such as atmosphere filters, oxygen regenerators, microgravity kitchens, and zero-g toilets needed to be developed. With this new equipment, astronauts could stay in space for months and also fly beyond low Earth orbit. NASA sent astronauts to the Moon and found that people could live and work there and make giant leaps in the low lunar gravity (see Figure 29).

Space stations such as the Russian Salyut series, Mir, and NASA's Skylab were launched and offered the possibility of studying the effect of prolonged microgravity on the human body. Cosmonauts stayed in orbit for over a year and found that, on return to Earth, they could readapt to gravity within weeks if they trained sufficiently while in space. The dream of human spaceflight, which had started scarcely a hundred years previously with some visions and little hardware, became a spectacular reality.

Spaceflight has become more and more routine, and few people know the names of the astronauts who now fly to the International Space

FIGURE 29   The giant Saturn V is the largest launcher ever built. It was used to send astronauts to the moon and to put NASA's first space station, Skylab, into orbit. [Photo: NASA]

Station. From a space race requiring daring test pilots as astronauts, human spaceflight has become something that is, in principle, possible for almost anyone. Short periods of weightlessness have no permanent effect on your health and the forces during a launch with a modern rocket are easy to cope with.

Astronauts, wherever they come from and whoever they are, all agree that a spaceflight is an astonishing adventure that forever changes your view on Earth and humanity. The experience appeals to many people, and technically we have made it possible for almost anyone to make a trip into orbit.

MAKING IT AFFORDABLE

It is mainly the high price of a flight into space that prohibits a large space tourism market and generally makes spaceflight the domain of professional astronauts. Dennis Tito and Mark Shuttleworth each spent an estimated $20 million for their launch with a Russian Soyuz rocket and capsule.

Flights costing tens of millions of dollars are only affordable for a very few multimillionaires. To enable the average person on the street to take a holiday in Earth orbit, the space tourism ticket prices will have to come down dramatically.

The problem of high spaceflight costs lies mostly with the launch vehicle. Most launchers used today are of the expendable throwaway type, meaning that nothing of the vehicle can be recovered for reuse. The stages of these rockets are dropped off along the way, splashing into the ocean or burning up in the atmosphere.

Expendable rockets have many benefits with respect to complicated reusable systems, especially from a military point of view: they cost less time and money to develop, you don't have to worry about how to maintain them for more than one flight, and you can use them to throw nuclear bombs at other countries.

Using the German V-2 rocket as a basis, Russia and America were able to build ever larger ballistic missiles in a relatively short time in the 1950s and 1960s. In the days of the Cold War military powers put a lot of money into expendable ballistic missile projects. The fact that the hardware and technology developed for these weapons could also be used for launching satellites and crewed capsules was a nice side benefit, but not the most important goal. Indeed, most space launchers today are direct descendants of intercontinental ballistic missiles (see Figure 30).

However, expendable launchers are inherently expensive in use: a medium sized launcher such as an Ariane 44L can put a 10,000-kilogram (22,000-pound) satellite into a low orbit, but in doing so it burns some 430 metric tons (960,000 pounds) of propellant and throws away 40 metric tons (90,000 pounds) of precious rocket hardware. Such a launch today would cost you on the order of $130 million, i.e. $13,000 per kilogram ($6,000 per pound) of satellite!

For human spaceflight, you would also need to fit an expensive capsule on top of your launcher, and as these capsules are typically not reusable, that makes a flight even more expensive.

The smaller the launcher, the worse the launch price per kilogram becomes due to the high "fixed" costs for launch control, launch pad

FIGURE 30   Rockets used to launch astronauts, such as the Saturn IB on the left of the picture, are directly derived from military ballistic missiles such as those on the right. [Photo: M.O. van Pelt]

FIGURE 31   *Although Europe's Ariane 5 is one of the most recent heavy launchers, it is still a fully expendable, rather expensive system. [Photo: ESA]*

security, etc., that are not a direct function of the size of the launcher. A small satellite of a couple of hundred kilograms will cost you some $20,000 per kilogram ($9,000 per pound). Larger rockets are relatively less expensive; an Ariane 5 launch costs "only" around $8,000 per kilogram ($3,600 per pound) for a low-orbiting satellite.

Still, even a really big launcher, such as the enormous Saturn V rocket that took Apollo astronauts to the Moon, could not make a tourist trip into orbit very cheaply. Let's assume that an average space tourist weighs 75 kilograms (170 pounds) and that the mass of the spacecraft and sufficient air, water, and food is about 1,200 kilograms per passenger. A Saturn V, with a 130,000 kilograms low-orbit payload, could then carry a spacecraft with about 100 passengers into orbit. In current year dollars, a Saturn V launch would cost some $2.4 billion. Assuming that we can half that by using more modern technologies and production processes, the ticket price would still be $12 million per person, not even counting the costs of the spacecraft. Not really in the range of the average holiday expenditure!

Market studies indicate that to enable a space tourism business to fly thousands of passengers a year, each ticket should not cost more than about $100,000. This is clearly not achievable with traditional, expendable launch vehicles (see Figure 31).

Using a brand new rocket and spacecraft for each launch is like buying a new car for each trip you make without the possibility of selling the old one! Making launches, and therefore space tourism, more affordable, requires reusable systems.

Ideally, you would like to have an airplane type of launch vehicle, one able to take you into orbit and bring you back on a routine basis. Modern airplanes are safe, reliable, can be used for thousands of flights, require relatively little maintenance, and are cheap to use; ticket prices for modern airliners are governed by the cost of the plane's fuel, not its production or maintenance.

THE DEVELOPMENT OF SPACEPLANES

Many early spaceflight visionaries assumed that astronauts would ride into orbit on board plane-type launch vehicles. They saw the spaceplane as a logical next step in aeronautical progress – one already started with the development of rocket-propelled airplanes.

Rocketplane development began with the German pre-war rocket sailplanes, which were followed by tests with winged versions of the V-2 and the rocket-powered Me-163 fighter during World War II. Germany even had plans for a rocket-propelled bomber spaceplane to reach America.

After the war, rocket-propelled planes like the Bell X-1 (the first plane to break the sound barrier), the Douglas Skyrocket, the Bell X-2, and the British SR.53 continued to push the limits of airplane and rocket technology.

Between 1959 and 1968 the successful series of fully reusable North American X-15 experimental rocket planes reached the edge of space, achieving a maximum altitude of 107 kilometers (66 miles) and the incredibly high velocity of Mach 6.7, about a quarter of the speed required to orbit a satellite. The brave X-15 Air Force pilots who flew over 80 kilometers (about 50 miles) high were even granted official astronaut wings (see Figure 32).

Much progress was also made on so-called lifting bodies – i.e. airplanes without wings that used the shape of their fuselage to create sufficient lift to fly. Lifting body spaceplanes have very rounded shapes, which helps to

FIGURE 32   *The X-15 of the 1960s reached the edge of space, was fully reusable and could have led to the orbital X-20 spaceplane. Sadly, the revolutionary X-20 was cancelled during development.* [*Photo: NASA*]

distribute the heat of atmospheric reentry over a large area and thereby keeps temperatures down. It is similar to the fact that when you hold a needle and a spoon in a flame, the point of a needle gets hotter and starts to glow earlier than the larger spoon surface. For this reasons the Space Shuttle was designed to have a very blunt nose and wing edges, even though it was not a lifting body.

Inspired by the successes of the X-15 project, many plans for true spaceplanes were drafted in both Europe and the US.

The US Air Force's crewed X-20 Dynasoar, intended as the successor of the X-15, was to be the first true spaceplane. Launched on top of an expendable Titan rocket it would enter orbit, perform its mission, reenter the atmosphere, and fly back to its launch site to be made ready for its next trip. The X-20 was to prepare the way for fast-response military spaceplanes that would be able to drop nuclear bombs on any target on Earth or fly spy cameras and radar systems over the Soviet Union.

Sadly, the X-20 project was cancelled in December 1963, less than a year before the plane's scheduled first flight. With Dynasoar the US could have had a partial reusable crewed launch vehicle as early as the mid-1960s, but at the time all funding was going into the simpler expendable launcher and capsule approach. Instead of following the path envisioned by von Braun and many other space experts of developing reusable rocket spaceplanes, converted military missiles and relatively simple capsules with parachutes were used to obtain rapid successes. Although this approach resulted in an astonishing pace of space spectaculars, it also meant that human spaceflight was not moving in the direction of efficiency and reusability. With the demise of the Dynasoar project and the various European projects never obtaining funding to move beyond the paper report phase, development work on potentially low-cost, reusable spaceplanes had virtually stopped.

It is difficult to say how much further spaceplane development would have progressed had Dynasoar flown, but the gained experience would certainly have positively influenced the currently uneconomic Space Shuttle design. We may have had an efficient, fully reusable spaceplane system in the 1980s or 1990s.

Only when the Moon race was over did the US take the idea of reusable spaceplanes from the drawers again. Many plans were drafted to develop single-stage reusable launch vehicles, but at the time this was not considered to be feasible by NASA and the US military. Instead, the Space Shuttle was created – a mix between traditional expendable multistage rocket technology and a reusable, winged spacecraft.

The Space Shuttle is an incredible machine. It is the first and only reusable spacecraft, a heavy lift launch vehicle that can also return heavy cargos, a spacecraft that, to date, has delivered three times more people to orbit than all other launchers combined, and has the most efficient rocket engines ever produced. However, it is also complex, labor-intensive, very expensive, and has a less than perfect safety record.

The Space Shuttle project had problems from its inception. Instead of a gradual next step after X-15 and X-20, the Shuttle represented a leap in size and technology. Moreover, the project suffered from an extremely long list of requirements: astronaut transporter, space station launcher, satellite retriever, orbiting laboratory, etc. It was supposed to do everything. The available funding was insufficient to develop a spaceplane that could do all of this and still be safe, reliable, and fully reusable. For instance, to save money an efficient, liquid propellant fly-back booster system was discarded in favor of simpler but far more dangerous and inefficient solid propellant boosters.

The Space Shuttle, which eventually flew for the first time in 1981 and will be operational until at least 2010, is only partly reusable. The two big white rocket boosters are retrieved by parachute and the Shuttle Orbiter glides back from orbit as an airplane, but the huge brown propellant tank is discarded. It burns up in the atmosphere because of the high velocity and the increasing friction with the air as it falls to Earth.

However, since the most expensive parts of the system, the Orbiter and the boosters, are reused, it was expected that the Space Shuttle would dramatically lower the costs of getting into orbit. It was supposed to operate as a kind of space truck, delivering communication satellites, space station modules, space telescopes, onboard laboratories and astronauts into space at an unprecedented low price and on a weekly basis.

It was not to be so; the costs of getting the Shuttle ready for flight proved to be tremendous and far outweighed the benefit of reusing parts of the system.

The checking and partial replacing of the 35,000 heat protective tiles on the Orbiter alone takes weeks and the complete post-landing overhaul of the Orbiter involves over 100,000 hours of work (i.e. it would take one person more than 100,000 hours to do all the work, or 100 people over 1,000 hours). The Orbiter then has to be mated with a new external propellant tank and the cleaned, refilled solid rocket boosters.

Next, the whole assembly is driven from the enormous Vehicle Assembly Building to the launch site on huge caterpillar Crawler trucks at a speed of less than 2 kilometers per hour (1.2 miles per hour). Once at the launch pad, propellants are loaded, electrical interfaces are established, and subsystems are checked and checked again, everything according to a schedule dictated by a long countdown checklist. All these operational activities make the Space Shuttle the most expensive launch vehicle at this moment.

As a successful orbital space tourism business, the Space Shuttle would be far too expensive. Assuming that a passenger module for 74 space tourists could be installed in the cargo bay of the Orbiter and that 12 flights per year could be made, the ticket price would be around $3.6 million. While there are probably a number of people willing to pay such an amount for a space flight experience, flying 74 of them at the same time would very shortly deplete the limited available market. Moreover, the Shuttle turnaround time (the period between landing and relaunch) of around 3 months would be unacceptable for a large-scale space tourism operation.

The original Space Shuttle requirements had demanded a low-cost, easy to process cargo plane for space, but development budget restrictions and technology limitations resulted in a costly and labor-intensive vehicle. The Challenger disaster and the tragedy with the Columbia proved that it is also not as safe as was expected.

In the Soviet Union developments on spaceplanes were started in the early 1960s. The "Spiral" concept was much more ambitious than the X-20 Dynasoar: the vehicle would have consisted of a reusable air-breathing launch aircraft carrying a two-stage expendable rocket with a small orbital spaceplane. A first version of the launch aircraft would have used kerosene fuel and accelerated to Mach 4 before launching the rocket/spaceplane combination at an altitude of about 23 kilometers (14 miles). A variant for the longer term would have reached Mach 6 and a height of 30 kilometers (19 miles).

After separation, the launch craft would fly back to its base while the spaceplane would go into orbit with the help of the expendable rocket stages. The orbital spaceplane was to be a lifting body with extendable wings for use during the subsonic (slower than Mach 1) part of the return flight.

The small spaceplane would have been able to house one cosmonaut-pilot, contained in an emergency escape capsule that could return to Earth independently even when ejected in orbit. The payload of the craft would have been relatively high, lowering the cost per kilogram of payload to orbit by a factor of 3 or more. The first flight of the complete system was planned for 1977 and a group of cosmonauts went into training to fly the test models and the actual spaceplane. Gherman Titov, Russia's second cosmonaut, was appointed to head the team.

It proved difficult to maintain support for the project from the higher levels in the Soviet hierarchy, because massive funding was needed, the technology was difficult to master and the long development program did not promise quick results. In 1970 Titov lost faith in the project and left the training group. In 1973 the team was disbanded and in 1976 the underfunded project was officially ended. Only the tests of an already built subsonic development vehicle were continued, which were to help to start a new shuttle project called "Buran" (Russian for "Snowstorm").

To test heat shield material for Buran, a series of small, uncrewed spaceplanes were launched on top of Kosmos rockets into suborbital trajectories and even into orbit. The orbital "BOR" vehicles used rocket engines to brake out of orbit and made a gliding reentry back to Earth. Parachutes were deployed at the last stage of the flight, after which the spacecraft splashed down in the ocean to be recovered by the Soviet Navy. The BOR vehicles were originally conceived as test models for the Spiral project, but were now employed to aid in the development of the Buran shuttle.

Although the Spiral and Buran projects were very secretive, pictures were made of a BOR-4 by an Australian Naval Reconnaissance aircraft while the spaceplane was being retrieved by a Soviet Navy vessel.

The final Buran spaceplane was an almost exact copy of the American Space Shuttle. It did not use its own rocket engines for launch, but was attached to the side of a giant Energia rocket. Buran made one uncrewed flight in 1988, before the financial troubles of the crumbling Soviet Union killed this project too. Buran was very similar to the Space Shuttle, and it would not have lowered launch costs with respect to expendable Russian launchers either.

REUSABILITY

The lessons learned from the Space Shuttle and Buran are that you need a truly reusable launcher that is easy to maintain if you really want to lower the price of launching things into space. Preferably, the system should involve only one single vehicle without expendable tanks or boosters that need to be retrieved and refurbished. Also, to avoid complicated and expensive transport, it should land at the same place from where it was launched.

Instead of thousands of fragile heat-resistant tiles, a limited number of easily replaceable metallic shingles should be employed. The propellants should be non-toxic, safe, and relatively easy to handle to avoid complicated tanking and propulsion system maintenance procedures. The rocket engines should last longer than those on the Space Shuttle, should require less maintenance, and should be easier to repair.

Ideally, the launcher would have some kind of combined rocket/jet engine that can use the oxygen in the atmosphere while flying at relatively low altitudes. This would mean less onboard propellant, smaller tanks and, therefore, a smaller, lighter vehicle.

To make maintenance more efficient, these vehicles of the future may carry computers and sensors that constantly check the health of all subsystems and components during flight. A readout from this system would make it easy to determine the type of maintenance and where it was required. Routine manual checks of the entire machine after each flight would be unnecessary.

Unfortunately, developing launchers that fit all or at least many of these requirements and are also able to bring a significant payload into orbit has proven to be extremely difficult. With the current rocket engine and materials technology, even expendable launchers are heavy and hardly powerful enough to make it into space without discarding parts of themselves along the way; rocket stages are jettisoned as their propellant tanks run empty. These stages fall in the ocean or crash on land and cannot be used again. A truly reusable launcher cannot benefit from such a simple staging system, but still must be able to take large satellites to orbit.

Another problem is that today's rocket engines can only be operated for 10 minutes or so before major maintenance activities or new motors are required. Jet engines, as used in modern airliners, last for months with no trouble.

The problem is illustrated by the fact that no one has ever been able to make supersonic airplanes truly profitable, even though such aircraft are simpler to develop than spaceplanes. (Concorde, for example, was operated commercially for many years, but never earned enough to compensate for the $12 billion investment in today's dollars, excluding engine development.) All future airliners now on the drawing board are subsonic, flying less than one-twenty-fifth of the speed required to fly into Earth orbit.

Work on reusable rockets for launching satellites has been continuing for many years now, without much success. NASA, the European Space Agency (ESA), and the Japanese Space Agency (JAXA) are designing test vehicles, performing design trade-offs and work on the development of new materials and rocket engines. However, until now only limited research projects with small, uncrewed experimental vehicles have been set up. Many of these have yet to make it beyond the paper studies phase, have been put on hold due to budget problems or project redefinitions, or have been cancelled completely long before their first test.

The NASA/McDonnell Douglas DC-X Delta Clipper was a small vehicle to demonstrate that a reusable rocket can take off vertically, hover, then land vertically back at the launch pad. The project team also wanted to show that with only a small crew such a vehicle could be quickly readied for relaunch after landing. The project was quite successful: the unpiloted vehicle made 12 test flights between 1993 and 1996, two of which were made within 26 hours. Sadly, on the last flight a landing leg failed to deploy and the craft tumbled over. The crash made the oxygen tank explode and the vehicle was destroyed.

NASA's X-34 was an unpiloted, experimental rocketplane developed by Orbital Sciences. It was to test spaceplane technology and, like the X-15, would have been dropped from a carrier airplane. However, NASA cancelled the project in 2001.

The most ambitious test project, NASA's and Lockheed Martin's X-33, involved a suborbital, single-stage reusable test vehicle. It was to launch vertically like a rocket, fly 15 times the speed of sound, then land horizontally like an airplane. Sadly, the project was scrubbed in 2001 because of major problems with the development of the engine and the lightweight hydrogen tanks. The cost of the project exceeded the $1.2 billion budget limit even before any test flights were made (see Figure 33).

Both the DC-X and the X-33 were regarded as models of larger reusable launch systems. Specifically, the X-33 was supposed to lead to the development of Lockheed Martin's "Venture Star" single-stage reusable spaceplane. However, it would have been very difficult to scale the test vehicle designs up to full size, operational launchers without gaining too much mass.

In March 2004, NASA successfully tested the X-43A, an almost 4 meter long unmanned scramjet airplane. The X-43A was an airbreathing vehicle that used oxygen from the atmosphere to achieve high speeds. Flying seven times the speed of sound, the velocity was sufficient to compress the air in the engine without the need for compressors, as in a regular jet engine. Spaceplanes with scramjets could use atmospheric oxygen during a large part of their flight to orbit, and thus needed to take less oxygen with them than normal rockets. The use of scramjets could thus result in smaller, more affordable spaceplanes.

Unfortunately, scramjets only work at hypersonic velocities; at speeds lower than Mach 5 the air is not sufficiently compressed for efficient combustion and propulsion. Moreover, at low altitudes the air density is

FIGURE 33   NASA's and Lockheed Martin's X-33 was to lead to the development of the larger Venture Star single-stage-to-orbit, fully reusable spaceplane. However, the X-33 proved too big a step and was cancelled before its first flight. [Photo: NASA/Lockheed Martin]

too high to fly Mach 5 or faster. The resulting pressures, forces, and temperatures would destroy a spaceplane. The X-43A had to be launched to high altitude and velocity on top of a converted Pegasus rocket, and the whole combination was dropped from a B-52 bomber. At high altitudes there is not enough air to enable a scramjet to work effectively. Scramjets are thus only part of the solution. Scramjet technology will probably be incorporated in so-called rocket-based combined cycle engines. Such engines would operate as a rocket for takeoff and to achieve hypersonic velocity, then switch to scramjet-mode to save the onboard oxygen. At high altitudes where there is no longer enough oxygen, the engine would once again work as a pure rocket.

Hidden in laboratories or more openly during test flights, the technologies needed for future spaceplanes are being developed. However, progress is rather slow. The world's space agencies seem to be in no hurry to develop reusable launch systems for a limited satellite launch market that does not require regular launches. It may prove to be cheaper to launch an expensive, expendable rocket occasionally than to invest huge amounts of money in reusable launcher technology.

The situation is a classical chicken-and-egg problem: as long as launches are expensive, the number of satellites and people to be launched each year remains small. Reusable launchers only become economical at high launch rates, so their development and operation are not justified for such a limited market. Expendable launchers therefore remain in use, launch costs remain high and, in turn, the satellite market stays small (see Figures 34 and 35).

Space tourism may help out, as it offers a clear, large market worth billions of dollars per year, where success depends on efficient, reusable vehicles that are making numerous flights. Reusable launchers developed for space tourism can reduce launch costs dramatically, enabling not only regular tourist flights but also cheap satellite launches. This would offer an enormous boost for the exploration of the solar system, the colonization of the planets, the construction of space factories and solar power-generating satellites, and other possibilities of which many have probably not yet been identified.

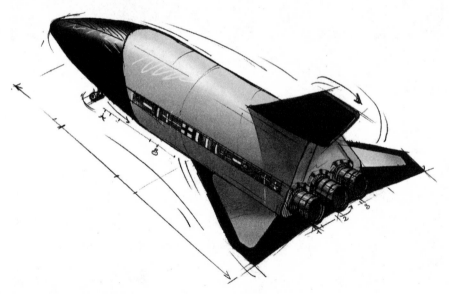

FIGURE 34 *There are many concepts for future reusable spaceplanes, but none has yet left the drawing board. [Photo: ESA]*

FIGURE 35   Each large space mission, human or robotic, requires a substantial mission control team of experts. This is an indication that spaceflight is still not really mature. [Photo: ESA]

FINANCING THE FUTURE

For a purely commercial enterprise the start-up costs for large-scale orbital space tourism have proven to be a big problem. Cost estimates for the development of a fully operational reusable launcher with modern technology are in the order of $10 billion. Although this is relatively low in comparison with the historical development costs of the Saturn V Moon rocket ($42 billion in today's dollars) or the Space Shuttle system (about $25 billion in today's dollars), it is much too high for a starting commercial space tourism business. Furthermore, the certification process for a space tourism launch vehicle (which currently does not exist) will be more elaborate and costly than for an uncrewed satellite launcher. Commercial airplanes are typically required to make about 1,000 test flights with the same aircraft before certification. At the current state of technology, spaceplanes don't even last 100 flights.

Even considering the most optimistic space tourism market predictions, it will take too long to recover such huge investments. Investors in the type of high-risk enterprises that the first space tourism companies will be, typically demand a high return on their investments within five years or so. If space tourism does not offer this, they will invest their money in other businesses.

Government funding in the development of reusable space tourism launchers could be a solution to this problem. Funding by a government organization is common in the launcher industry; for instance, the development of the European Ariane launchers is funded by ESA, while the commercial operation and marketing is under the responsibility of a private enterprise named Arianespace.

Governments may support space tourism directly, or develop efficient reusable vehicles for launching satellites, which can then also be used by commercial space tourism enterprises. This is a likely scenario, as space agencies are very hesitant to endanger the lives of non-professional astronauts in any way. A single serious accident with a space tourist launch vehicle could endanger a whole government space program because of the political implications. Commercial companies like airlines are usually able to survive such problems since they are not directly related to politicians' public images. Launching satellites can also be used as a way to make money in a development or certification phase, during which no space tourists can be allowed on board.

Another way of financing the development of orbital spaceplanes could be a step-wise approach, as suggested by David Ashford of Bristol Spaceplanes Limited. Starting with a small, suborbital X-Prize vehicle, a company could make money that can be invested in the development of a larger two-stage spaceplane consisting of a large hypersonic carrier airplane with a smaller shuttle on top. At high altitude and velocity, the upper vehicle would be released from its mothership and launched into orbit. Systems consisting of two stages are less economical in operation than single-stage spaceplanes, but less difficult and therefore less expensive to develop.

From there on, larger two-stage vehicles and eventually single-stage spaceplanes could be developed. Just as normal airplanes, these advanced spaceplanes would use air for their engines while flying at low altitudes to save on the amount of propellant that needs to be carried along.

Following this approach, development costs and risks are spread out while money can be earned long before the fully operational orbital vehicle is flying. It also allows the gradual build-up of a space tourism market.

Whether tickets to orbit for a couple of thousand dollars will become available to us or not depends on the success of spaceplane technology developments and the willingness of investors, companies and governments to put considerable amounts of money into reusable launcher programs. As the saying goes: No Bucks, No Buck Rogers.

NUCLEAR ROCKETS, LASER CRAFT AND SPACE ELEVATORS

At this moment, the only operational means of astronaut transportation are the Space Shuttle and the Russian Soyuz. The Shuttle is more than 20 years old, the Soyuz has even been in use for nearly 40 years now. The next three generations of launch vehicles will almost certainly be based on conventional rocket engines again, maybe in combination with the partial use of airbreathing engines. It seems that astronauts, both professionals and tourists, will continue to go into orbit on board rocket-propelled vehicles for at least the first half of the twenty-first century.

However, there are limits to where you can get in a reasonable time and at a reasonable price with conventional rockets. No matter how much we improve chemical rocket engines in the future, they will always require enormous amounts of propellants and therefore large launch vehicles to put relatively small loads into space. There are still some ways left to make these engines more economical, but we should only expect marginal improvements in performance.

This becomes very apparent in designing human space missions to Mars, for example; spaceships based on chemical propulsion that need to take all their propellant are hugely expensive, mostly expendable, and rather unpractical monsters.

But are rockets based on the combustion of propellants the only means of space transportation available to us? The fact that all near-future launch vehicle concepts are based on conventional rocket technology does not mean that no one is considering anything else. Not all of the possibilities are suitable for launching people, however, and most will not be ready for use in the near future.

Light gas guns, electromagnetic rails, and coil guns are concepts for launching payloads into space with sophisticated, extremely powerful cannons. Just like Jules Verne's giant Columbiad, the problem is that the launched projectile needs to attain an extremely high velocity over a very short distance: the length of the gun. This means accelerations of over 1,000 g, which is clearly not healthy for anything that is not made of extremely dense, solid materials. Human beings would be smashed to a thin film on the bottom of the capsule within a fraction of a second!

However, electromagnetic rails with lower accelerations could assist the launch of more conventional rocket vehicles so that they would require less propellant to get into orbit.

Nuclear fission or fusion rockets are advanced concepts that could be used to launch astronauts. At the end of the 1950s American scientists working on nuclear weapons hatched a plan to propel a giant spaceship with atomic bombs. This "Orion" spaceship, truly a space*ship* rather than a space*craft*, was supposed to fly astronauts to the Moon, Mars, and even the moons of Jupiter and Saturn in the late 1960s and early 1970s.

With intervals of less than a second, Orion would shoot out hundreds of small nuclear bombs at the back. At some distance behind the ship these bombs would explode, propelling Orion by hitting a special bumper shield, the so-called "pusher plate," with bomb debris. To prevent a rather bumpy ride, the pusher plate was connected to the rest of the ship by a set of giant shock absorbers.

After initial interest by the Air Force and NASA, the project soon lost support because of the huge leap in technology that would be required. Moreover, the public and politicians became increasingly sensitive to exploding nuclear weapons in the atmosphere; launching Orions from the Earth's surface would have resulted in large amounts of deadly radioactive fallout.

The designers suggested an alternative plan to launch smaller Orions with Saturn V rockets into Earth orbit and have them start their trip to Mars or beyond from there. However, at the time NASA was already working on another nuclear propulsion concept called NERVA, and thought Orion was a step beyond that, for later in the future. The Air Force was inclined to partly budget the development, but not without NASA chipping in. Moreover, the Air Force could not come up with a military role for Orion that would justify the spending of billions of dollars. The project was mothballed, but recent NASA interest suggests that the concept may be resurrected in the near future.

A more modest US nuclear fission propulsion project was NERVA. During the 1960s and 1970s experiments were performed on NERVA nuclear fission rockets in the United States, for possible use in crewed spacecraft to be sent to Mars.

Energy from a nuclear generator was used to heat hydrogen gas to extremely high temperatures. The hydrogen was then expelled through a conventional rocket nozzle. NERVA's concept was proven to work and the technology is quite mature, but the work has been all but abandoned.

FIGURE 36 *The nuclear NERVA rocket stage in a museum in Huntsville, Alabama. Nuclear propulsion was never used but the technology is mature and could be applied to send people to Mars and further. [Photo: M.O. van Pelt]*

Only recently has NASA proclaimed renewed interest in the technology for application in interplanetary spacecraft (see Figure 36).

For use in a crewed spaceship, a nuclear fission engine would have to be separated from the crew cabin by some distance and a radiation shield installed between the astronauts and the reactor.

It is, however, highly unlikely that nuclear fission rockets will ever be launched within the Earth's atmosphere because of the radiation pollution risks involved.

Nuclear fusion is based on merging atoms instead of breaking them apart as in a fission process. Fusion rockets could be a lot safer as they generate much less harmful radiation and radioactive material, but the technology development is in a very early stage. No one has yet even succeeded in maintaining a nuclear fusion process that provided more energy than the amount used to sustain the reaction.

An exotic implementation of fusion propulsion is in a ramjet that scoops up the hydrogen gas in interstellar space and feeds it to a fusion reactor to produce thrust. Unfortunately, hydrogen is very sparse between the stars, and in order to produce enough power, the scoop would need to be thousands of kilometers across.

Another very interesting technology that has just entered the experimental stage is laser propulsion. The idea is to heat propellant on board a vehicle with a ground-based laser and expel the resulting heated gasses through a rocket nozzle. The vehicle would require much less propellant to reach orbit and all the energy-generating equipment could remain on Earth. The downside is that the laser has to be very powerful, especially for launching a heavy crewed spacecraft.

In the US, experiments are performed with a small projectile that focuses incoming laser light on the air under it. The laser is powerful enough to make the air explode and thereby boost the metal object up into the air. Altitudes of several tens of meters have been achieved so far.

Whether laser technology can ever be used to launch people into orbit depends on the progress made in the development of large, extremely powerful lasers. Those currently under development to shoot down incoming ballistic missiles may some day find a more peaceful purpose.

A related means of transportation are solar sails, which use the tiny pressure of light hitting a reflecting surface to propel large, ultra-light spacecraft with enormous solar sails made of extremely thin foil. Obviously, these cannot be used to launch anything from Earth, but they could be highly efficient for travel to the Moon or between planets. The small thrusts involved would mean that passengers would have to be very, very patient, however.

The use of large lasers on Earth to push the spacecraft could enhance the effectiveness of solar sails, although a laser loses its usefulness over long distances because the beam diverges as a function of the distance it travels (in fact, proportional to the distance squared).

A very different way to get to orbit, and one that a recent NASA study says may be feasible in 50 years or so, is the space elevator. According to the NASA plan, a cable of an incredibly strong material would be used to connect an orbital space station directly with the Earth's surface. The center of mass of the system would have to be positioned in Geostationary Earth Orbit, where its orbital rotation would be 24 hours and it would stay over the same point above the equator as the Earth rotates around its axis. In any other orbit the system would move too fast or too slow and break the cable. A space elevator would make a physical connection between Earth and space, similar to a bridge linking two places across a river.

Electromagnetically suspended elevators or mechanical climbers could shuttle between the surface and the station without the need for any rocket propulsion. The cost of bringing anything into orbit could be lower than $10 per kilogram, based on the required energy and its price, putting an average person with luggage into orbit for around $1,000.

If a space station could be put in the elevator's system center of mass, it would be almost 36,000 kilometers (22,400 miles) high. Travel time could therefore be rather long; more than a day for an electromagnetically propelled elevator and about a week for a mechanical climber. The elevator shuttles would then have to include sleeping facilities, a kitchen, toilets, etc., basically acting as small space stations on their own.

Going up, the elevator would, at a certain point, reach a maximum speed and not accelerate further. From then on you would very slowly feel yourself getting lighter and lighter, until you reach the space station where everything would be weightless.

The orbital station could be made very large, as construction materials could be transported easily and cheaply with the elevators once the cable system had been completed. It could include telecommunications equipment, a satellite launch facility, astronomical observatories and, for tourists, large hotels with magnificent views, 0-*g* swimming pools and other sports facilities, casinos (with three-dimensional roulette) and observatories with large telescopes to gaze down at the Earth.

The feasibility of this concept depends mostly on the development of materials that can withstand the incredible forces on the cable. Steel is too heavy and not nearly strong enough. Recently developed nanotube materials, based on cylindrical molecules of carbon, may do the job. Unfortunately, the longest nanotube made so far is just a few feet long, so we are not yet able to produce long cables from that source.

Once in low Earth orbit, a system of freely rotating cables or "tethers" could be used to transfer spacecraft to higher orbits. Such a system would consist of a very long tether, 100 kilometers or more in length, with a ballast at one end and a sort of spacecraft catching/docking device at the other. Itself in orbit around the Earth, the cable would be spinning like a sling, with its center of rotation near the ballast mass. A spacecraft in a lower orbit could hook itself onto the docking end of the tether and let itself be swung into a higher orbit. With a series of these tethers, spacecraft starting just above the atmosphere could be shot all the way into an interplanetary orbit.

A spacecraft accelerated in this way does so by stealing momentum from the tether. By transferring energy to the spacecraft, the tether's own orbit is lowered. Eventually, it would fall back down to Earth if nothing was done to reboost it into a higher orbit. This could be achieved by lining the long cable with metal wire and passing an electric current through it. The interaction of the magnetic field generated by the electric wire with the Earth's own magnetic field results in a small but steady force that could push the tether back up. The power required could be generated by solar panels. This concept of generating a force with a cable and a current has been tested during a Space Shuttle mission, and it worked.

None of the above means of transportation is going to revolutionize human space flight in the very near future, but just like the replacement of propellers by jet engines made commercial mass transport by air possible, so one day normal rocket engines may be replaced by technology that opens up the entire solar system to everyone. Taking the family car out for a holiday on the Moon may not be impossible for ever.

7

IN ORBIT

THE pilot advises everyone to keep the safety belts on as he slowly rotates the spaceplane. It is better to adjust yourself to the microgravity environment while you are sitting in a chair, and there are still some orbit-adjustment burns to be performed.

Out of the window the stars show themselves to be much brighter than you have ever seen them anywhere on Earth. They don't twinkle because there is no atmosphere to distort their light, but their colors are much more distinct; some stars are vividly red or blue. You also see the Moon, but it doesn't look any different as you are less than 250 kilometers (155 miles) above the Earth's surface and the Moon is 384,000 kilometers (239,000 miles) away.

Now the Earth slowly slides into your window. The planet almost fills your entire field of view as you push your head to the glass as close as possible. It is an absolutely breathtaking view no movie or picture could ever hope to match. The oceans are so much bluer than you imagined! The bright reflection of the Sun shining in the water moves with you as you glide through space. It really is a blue planet, for the most part covered in oceans, seas and lakes. Although you have always known this, actually seeing it comes somewhat as a mind leap. People always seem to regard the land they inhabit as the most important part of the planet, but this blue globe should have been called "Water" instead of "Earth".

As you move further, the open water is replaced by a cloud-covered area. You see hair-like Cirrus clouds and the gray-brown hues of the land under them. The shadow of the clouds that is projected on the surface adds a distinct three-dimensional feel to the view. If you'd remember to watch your computer screen you could find out which country it is, but you are so absorbed by the view that you forget all about that.

A huge area of snow-covered mountains comes into view, looking like crumpled sheets of paper with mighty but now tiny looking glaciers carving their way through. The snow does not look perfectly white; all colors are tinted blue due to the light scattered by the atmosphere.

Two minutes later you recognize the Mediterranean, with the boot of Italy clearly peeking under the clouds. Slowly you drift southeast, toward the north of Africa and Saudi Arabia. The color of the surface turns from deep blue to a vivid ochre as you cross the desert, lined with rows of what must be huge sand dunes and rippling patterns caused by the wind.

The Earth is quite bright, all stars within 10 degrees of the rim of the planet are drowned in the reflected sunlight. You note what looks like a thin line of hazy blue and realize it is the atmosphere. On Earth the air looks never ending and seems to form a thick cover over the surface, but from orbit this vital, protective layer between breathing animals, plants and people and the hostility of space seems to be nearly nonexistent. If the Earth was the size of an orange, the atmosphere would be about 0.5 millimeter (0.02 inch) thick, much less than the fruit's skin. Life can only exist in one-tenth of that. The realization makes the planet look much more fragile than you ever envisioned it before.

A sudden "thump," followed by a continuous faint buzzing sound, disturbs your thoughts. The automatic pilot just initiated an orbital adjustment burn with the smaller rocket engines. You feel a slight acceleration as your back is softly pressed into the chair once more.

From the window you can now see that you are entering the night side of the planet. A narrow band of shadow marks the border between morning and night-time. As you cross New Zealand, lights of cities and towns peek through the darkness, the first obvious signs of human presence you have seen from orbit so far. When the cabin lights are turned off and your eyes adjust to the darkness, you start to distinguish the roads and smaller towns that form a luminous web over the black land.

Sudden dots of spooky white light erupt through the dark clouds that are hugging the surface, illuminating them from within. Thanks to your recent Earth observation training you realize you are witnessing an impressive display of night-side lightning.

A glowing false horizon of soft red and green light is apparent above the limb of the dark Earth. You realize it is the so-called "airglow" layer. During the day, the Sun's ultraviolet radiation hits the upper atmosphere, breaking apart molecules and knocking electrons from their atoms at an altitude of about 80 kilometers (50 miles). On the night side the molecules and atoms reform, emitting light in the process.

Looking through a window at one of the spaceplane's wings, you notice a faint glow there as well, just above its surface. As the spacecraft circles the Earth at tremendous velocity, it continually slams into the atomic oxygen present at this high altitude. The collisions make the atoms radiate a faint orange light.

EARTH ORBIT

The whining sound in the background stops as the spaceplane's engines are shut off. The pilot announces that you have now entered a stable orbit and asks you to carefully take off your helmet and boots and store them under the chairs. You put on the socks you will be wearing while you are in orbit.

As you are now less likely to hurt someone accidentally, you unstrap yourself and get out of the seat. To your amazement, you are hovering several centimeters above your chair. You find yourself naturally assuming a fetus position with your legs bent and retracted under you, just as you would do floating in a swimming pool. With your arms you are able to ease your way around the cabin. You soon notice that legs are not required to move around and are better to be kept out of the way.

As you bump into a fellow crew member, both of you spin around and move away from each other. Luckily the walls of the cabin are padded with soft materials and there are no sharp edges anywhere. You find yourself upside down and it takes a second for your orientation to grasp the situation. From your new viewpoint everything looks unfamiliar but the colors of the cabin – light blue for the ceiling and upper parts of the walls, dark blue for the lower parts – soon help you to realign yourself.

The sudden move has given you a strange feeling in your stomach, while your head suddenly seems to be a balloon filled with water, but nothing serious enough to start you looking for the vomit bags. Some of your colleagues appear to experience more serious problems and have started to ask the pilots for the Intramuscular Promethazine injections to alleviate the distress of space sickness.

Moving around in microgravity is wonderful. Just by softly pushing yourself you can float from one place to another. You quickly find that you have to plan your moves in advance, because while in mid-air, out of reach of the walls or any fixed object, there is no way to alter your trajectory. The cabin of the spaceplane is relatively small, but when you later visit the space hotel you will have to take more care.

There is enough room to tumble around, pull your legs in to make multiple somersaults, and bounce between the floor and the ceiling. In the old Mercury capsules, with less than 2 cubic meters (21.5 cubic feet) allowed for each astronaut, there was no room to move. One astronaut noted that you didn't enter a Mercury capsule; you wore it! The old Vostoks were just as compact. The Apollo Command Module that was used to fly to the Moon was slightly more spacious with nearly 3 cubic meters per person, but real, enjoyable space only arrived with the Space Shuttle (10 cubic meters, or 108 cubic feet, per astronaut) and the first Salyut and Skylab space stations (more than 50 cubic meters, or 538 cubic feet, per person). The *Jupiter*, with about 4.5 cubic meters (48 cubic feet) per crew member, is not very spacious but is comfortable enough. For real freedom of movement you will have to wait until you have docked with the space hotel.

As you are moving around the Earth every 90 minutes, dawn is already drawing near. First, a bright red strip marks the edge of the darkened planet. Soon an additional layer of brilliant orange appears on top of it, followed by a light blue strip that turns darker violet and finally flows into the blackness of space. As the Sun rises above the atmosphere, it seems that a wave of molten copper flows over the surface of the clouds. It is a magnificent display of color you will be able to witness more than 40 times in the coming days.

While more and more of the Earth is uncovered by the sunlight, a huge swirling storm comes into view as an enormous white snail shell projected on a dark blue background. Although you cannot actually see the clouds turn in circles, you know that the air under there is moving at velocities well over a 100 kilometers per hour (62 miles per hour). It is better to be up here.

At other places too the clouds create interesting patterns: cotton rolls, checkered patterns or fluffy islands resembling milk poured into a cup of coffee. There seems to be an infinite number of shapes.

Now your home country comes into view. Quickly you look for points of recognition as you try to locate your home town. It is not easy, because other than on a geographical map, there are no color codes, states, provinces or borders to be seen. Nor is the surface under you neatly positioned in a north–south direction.

Stubbornly you look for obvious points of reference, such as rivers and coastal features, until you finally find your home town. It is amazing to realize that virtually everything you know and love is down there: family, friends, pets, the office, the corner grocery store, all within the frame of a single window in a spaceplane. But all too soon it disappears from view, and you check the computer to see when your country will next come into view.

The orbit of the spaceplane has been chosen in such a way that most passengers will be able to view their home country from space. The inclination of your orbit is 52 degrees, the angle at which it intersects the equator. This means that you will cross every geographical latitude between 52 degrees North and 52 degrees South. Because the Earth turns under your orbit, everything in a band defined by the line Montreal–London–Berlin in the north and the latitude through Terra del Fuego in Argentina in the south will pass within your view during the time you are in orbit. At least five times, in fact.

However, space tourists from Scandinavia, Alaska, Greenland and the north of Canada and Russia will not pass over their homes. The inclination of the orbit could be increased to permit such views, but only at the price of a less optimal launch trajectory that constricts the spaceplane to carry less mass into orbit. This would mean that fewer passengers could be taken on board and ticket prices would be increased. From the spaceplane's performance point of view, the most optimal orbit would be one with an inclination equal to the latitude of the launch site: 28 degrees. However, since that would prevent many passengers from seeing their cities, towns or villages from space, an orbit with a 52-degree inclination is chosen as a compromise between maximum view and maximum spaceplane performance.

Below, a long river twists itself through the land, ending in a fractal-shaped delta where the water flows into the sea. The water near the estuary is colored brown, as mud carried by the river mixes with the seawater. Similar discoloring can be seen near certain cities on the coast, where waste is disposed into the water.

Forest fires throw thick clouds of black ash and smoke high into the atmosphere. The haze starts in a small spot but spreads out as it is carried by the wind over hundreds of kilometers.

You note how easily and clearly these forms of pollution can be spotted from space along with the large parts of the Earth that they affect. Together with your new appreciation of the thinness and fragility of the atmosphere, this revelation disturbs you quite a bit. You wish they would send world leaders up here before the start of important environmental protection summits.

ACTIVITIES IN MICROGRAVITY

Time to eat. You return to your seat and loosely strap yourself in again. The pilots have taken the plates with dehydrated lunches from the storage space in the tiny galley, which is located in the aft end of the cabin's upper level. They inject some of the packages with hot water to make soup or with cold water for lemonade or milk. Dehydrated food requires less valuable space and is easy to store for long periods, but there is also fresh fruit and some soft cookies that do not crumble.

The copilot brings you your plate, filled with plastic cups and small bags containing cold and warm food. Hungrily you open the first one; it has a label that says it contains pieces of chicken. You try to smell the food, but you have almost to stick your nose into it to detect any aroma since the warm air coming from the dishes does not rise up to your nose in microgravity. The next package contains thick yogurt with corn flakes. This sticks so well to your spoon that it does not float away. The bag with soup must be kneaded to ensure that all the powder has dissolved. You then remove the cap from the narrow opening and squeeze the soup directly in your mouth.

The candies are the most fun. Another passenger throws them at you and you try to catch them in your mouth. Throwing them requires some training: because of the lack of gravity they do not follow a parabolic trajectory but move in a straight line. At first you tend to aim too high, but you soon get used to it and are able to direct them in slow motion toward someone on the other side of the cabin.

When you try to suspend a candy motionless in the air, you notice that it soon starts to drift by the airflow from the climate conditioning. Without the continuous movement and filtering of the air, your breathing could create bubbles of carbon dioxide around your head and cause a lack of oxygen. This is especially dangerous during sleep, when you are motionless.

Next, you squeeze some lemonade from its plastic bag and admire the perfect spherical shape it assumes. In microgravity, the attraction between the atoms is the dominant force acting on a liquid. The atoms try to get as close as possible to each other, and the most optimal shape for this is obviously a ball. When you blow at the little sphere, it deforms and vibrates until it finds its equilibrium

FIGURE 37   *Even a water bubble becomes something magical in microgravity. [Photo: NASA]*

shape again, but it also floats away. Quickly you purse your lips and suck the lemonade into your mouth before it hits the chair in front of you (see Figure 37).

After the lunch, you clean your fork and spoon with wet wipes containing a strong disinfectant and store them for the next meal. Some microbes multiply and spread very quickly in a confined microgravity environment, such as the spaceplane cabin, so it is very important the keep everything clean.

Smoking is not allowed on board the spaceplane, primarily because of the problems that cigarette smoke would cause to the life-support system that recycles the cabin air. Floating ash and glowing pieces of tobacco would also be potentially dangerous for the crew and sensitive equipment. In fact, the air on board the spaceplane is cleaner than on Earth; in particular, people suffering from hay fever enjoy the pollen-free atmosphere.

The air pressure inside the cabin is the same as in a pressurized airliner, and is only a little less than on Earth at sea level. Also the composition is similar, with 80 percent nitrogen and 20 percent oxygen. The environmental control system circulates the air through filters to remove carbon dioxide and other impurities. Excess moisture is also removed to keep the humidity at a comfortable level.

After lunch you take pictures and movies with the digital camera. Mountains, clouds, rainforests, wetlands, river deltas – there is so much to see! Each time you take a picture the computer automatically registers the geographical position of the spaceplane at that moment, which enables you, at a later stage, to find the places you put on film.

As you dictate notes to the computer, it automatically converts your spoken words into writing, which can later be edited using the small keyboard attached to your seat. In the past US astronauts used special "space pens", which had pressurized cartridges to allow the ink to flow out of the tip in microgravity. The Russians had a simpler solution: they just used a pencil. ⇐

It is soon your time for making a phone call to Earth. You have a few minutes to talk to your family, using a video telephone that connects to a geostationary communications satellite in a much higher orbit. You find it very difficult to express the wonder and excitement you are experiencing; as many astronauts before you have discovered, there do not appear to be appropriate words to describe the view.

Next, it's time to have a look at Anita, the spider that has accidentally been taken on board. When it was found in a corner of the cabin after launch, it was quickly adopted by the crew. The arachnid already seems to have adapted to its new environment and has built a complete web, even though it is not as regularly shaped as it would have been on Earth. Although hard to see, spider web threads made in microgravity are thinner than usual; apparently, spiders use the sense of their own weight on the web to determine the amount of silk they need to spin.

After a whole afternoon and evening of more sightseeing and a dinner, the clock indicates that it is time for some sleep. With a new dawn every half hour, there is really no way of telling the time other than by a look at your watch or the timer on your computer screen.

You have to wait your turn to go to the small bathroom at the lower cabin level. There is no shower, since that would consume too much water and space and would be rather complicated: a powerful suction system would be required to remove the floating water droplets and you would have to wear an oxygen mask to prevent suffocating in the watery mist inside the enclosed cabin. Instead, you can take a warm sponge bath. This does not require much water as in microgravity the drops easily stick to your body and don't stream down (see Figure 38).

Other than a shower, the small onboard bathroom includes everything you would find in a bathroom on Earth, but many things are a little different. Towels need to be clipped to the wall and you must put your feet inside loops on the ground if you wish to stay in front of the mirror.

Washing your hands can be done by sticking them through two holes in a clear plastic bubble resembling an upside-down salad bowl. It is fitted with a

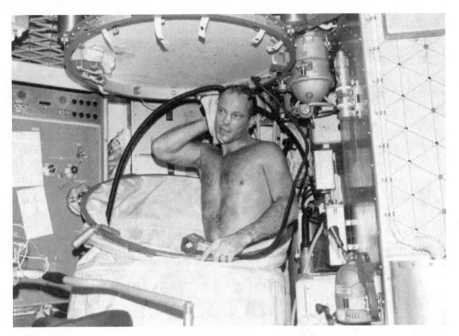

FIGURE 38 *NASA's Skylab space station had a shower, but it was hard to use: in microgravity water doesn't fall down but forms thousands of tiny, floating drops. The shower therefore had to be completely sealed during use, and afterwards the water had to be painstakingly removed. [Photo: NASA]*

water tap and a soap dispenser. The water is kept inside the bubble by gentle suction; even this small everyday activity is not straightforward in orbit. After brushing your teeth, you can't just take in some water and spit it into a sink, owing to the free-flying mess it would cause. Instead, you use edible toothpaste that you swallow together with the water.

There is also an ordinary looking razor, but it is attached to a flexible pipe through which the cut-off hairs are immediately sucked away. Alternatively, conventional shaving cream and a safety razor can be used, together with a disposable towel for cleaning your face.

The water used for washing is not reused on this spaceplane, nor is any other waste; because of the short duration of the flight it is more economic to bring fresh supplies than a heavy recycling system. On space stations, however, much water is recycled for drinking and cleaning – even urine and the condensation on the walls caused by sweating astronauts!

The amount of garbage accumulated during the three-day trip is negligible. On long-duration missions on board space stations it is a different story, however. The crews of the first Russian Salyut space station simply ejected their garbage bags through the airlock. But instead of moving away, the bags stayed close by, surrounding the Salyut with a cloud of trash that could even be seen from Earth as a collection of small dots surrounding the brighter dot of light of the station itself.

From then on the garbage was put on board the uncrewed spacecraft that were periodically sent to space stations to bring supplies, after their cargo was unloaded. The disposable capsules were subsequently sent back into the atmosphere, where they burned up completely.

After some time in front of the bathroom mirror, you decide that it is impossible to put your hair in order since it always floats upward. You return to the window at your seat to have another look at the Earth before you go to sleep.

Before you can enjoy your first night's rest in space, you have to change your seat into a bed by rotating the chair over 90 degrees (as not to put your head on the chair of your neighbor behind you) and putting the back into a horizontal position. Next, you roll out a thin sleeping bag, tie it to the chair, crawl inside and pull the straps around you. The bag is completely closed, with only your face appearing. You ensure that the straps around your head are well fastened, because if your head fell into the sleeping bag while you were asleep, the lack of air circulation could result in carbon dioxide pooling around your face. This could lead to suffocation through oxygen deprivation, with a monumental headache as the least serious of possible results. A soft restraint keeps your head down against the pillow and prevents you from waking up with a sore neck.

You are not actually lying down; if it were not for the straps, you would float away, propelled by your own breath! In fact, you could just as easily have strapped yourself to a wall or to the ceiling. However, for orientation purposes and to keep the passageways clear it is better that everyone sleeps on his or her own chair. Soon the lights in the cabin are dimmed and you close the cover on your window, after a last look down at planet Earth.

There is some noise inside the spaceplane, caused by the air fans of the life-support system. Although these are quite powerful and are able to recycle the air in the entire cabin every seven minutes, they have been designed to be as quiet as possible. The noise levels in the early spacecraft and space stations could be quite high, but here it is much the same as in a conventional airliner and the sounds do not disturb you. Earplugs are available for those who are especially sensitive to noise.

During the night, you suddenly wake up a few times because you feel as if you are falling. This is a natural instinctive reaction, as your body interprets the feeling of weightlessness as a sensation of falling down.

As you wake up in the morning, you are surprised to see a couple of hands floating in front of you. After some confusion, you realize they are yours! Better to tuck them inside the sleeping bag next time.

After breakfast you spend more time Earth-gazing. As the spaceplane passes over Egypt, you are able to discern the triangular shadows cast by the Pyramids, some of the most remarkable human-made structures that can be seen from space by the naked eye. Off the coast of Dubai you spot two artificially created islands shaped like palm trees. Artificial structures are usually easy to recognize because of their unnaturally straight lines.

Looking into the reflection of the Sun, land looks black while water appears to be transformed into liquid gold. Seen from this angle, all sorts of complex current patterns appear in the ocean. The turbulence moves too slowly to be detected by the boats crossing them, but from your unique viewpoint the disturbances become clearly visible.

You note that Anita has built a new web, as the earlier one was damaged when a crew member accidentally bumped into it. Remarkably, the spider seems to have adapted further to microgravity, as her new creation looks much more symmetric and regular than her earlier web.

THE "FREEDOM FLYER"

The pilot informs you that the plane is nearing the space hotel and advises everyone to buckle up in their seats and prepare for docking.

The small reaction control thrusters of the spaceplane are fired in short bursts to adjust the orbit and bring the vehicle slowly closer to the space hotel.

Rendezvous operations in orbit are quite complex: acceleration in any direction not only changes the velocity of a spacecraft, but also its orbit. Acceleration means moving up into a higher orbit, deceleration results in a lower orbit. However, a lower orbit means going around the Earth in less time, so with respect to the surface you are actually going faster when the spaceplane brakes! Luckily the pilots are assisted by very accurate guidance systems that automatically put the spaceplane in the same orbit as the space hotel and gradually align the upper hatch with the hotel's docking port.

The spaceplane, and especially the space hotel, are relatively fragile and as the orbital velocities involved are very high, the whole rendezvous and docking operation is performed in slow motion. Accidental bumps can easily wreck a spacecraft, so the fast maneuvers seen in most science fiction movies are out of the question (see Figure 39).

FIGURE 39 *A Soyuz spacecraft floats over the blue Earth as it approaches to dock with the International Space Station. [Photo: ESA]*

The spaceplane slowly rolls to correct its attitude and majestically the space hotel glides into view. It looks like a large white can topped by a golden box with two emerging wings of solar panels. It is basically a small space station, with the can being the pressurized crewed module and the box housing most of the life-support, power, communications, data-handling, thermal and attitude control equipment.

In spite of its large size, the entire hotel was launched in one piece because the pressurized module is an inflatable structure. It was put in orbit in a folded configuration, then inflated. The walls of the module contain a material that hardens after inflation, creating a strong structure that does not collapse if punctured by a micrometeorite.

The station is popularly called the "Space Hotel," although its official name is "Freedom Flyer" and it should not really be called a hotel. Apart from the view, it does not measure up to any Earthbound tourist resort; the interior is quite Spartan as the module is merely intended to offer more room for the space tourists than that offered by the spaceplane.

Real hotels, where guests will be able to stay for a few days, will be put in orbit soon. One will even house a casino; since space is not owned by any country, gambling legislation is nonexistent in Earth orbit. Nevertheless, most companies that organize space tourism flights and hotel arrangements have decided that the law of their home country on Earth is in effect on their spacecraft.

Slowly the spaceplane glides nearer to the small station. Small laserbeams reflecting off reference mirrors on the space hotel are used to align the huge spaceplane with an accuracy of less than 1 centimeter. You feel a sudden bump as the docking cone hits the space hotel's adapter and glides home, almost immediately followed by a sudden stop and a loud clang as the docking is completed.

It takes a few minutes to equalize the atmospheric pressures in the spaceplane and the space hotel, but then the pilot opens the hatch and floats into the huge inner volume of the "Freedom Flyer". He first does an elaborate check on the systems of the station to ensure that everything is working correctly and that it is safe for the passengers to enter.

The space hotel is in fact a very simple space station. For life support, for instance, it depends on the spaceplane's systems because the "Freedom Flyer" is only used when a spaceplane is docked to it. Fans circulate the air from the space hotel to the spaceplane and back. There is an independent life-support system on board the station, but that is only for use during emergencies – for instance, if a crippled spacecraft with a malfunctioning life-support system is docked and the passengers have to shelter inside the "Freedom Flyer." A stranded crew can safely survive for a few days, leaving ample time for another spaceplane to be sent from Earth to rescue them. As the evacuation vehicle can dock to a second hatch at the other side of the station, the damaged spaceplane would not have to be disconnected.

To ensure that there is enough room to play around in, the passengers are admitted to the space hotel in small groups of four persons. There is one day minus sleeping time available before the spaceplane will undock, so each group can stay for about two hours inside the "Freedom Flyer" module.

You are in the first group and eagerly push yourself in the direction of the docking tunnel. Entering the station, you are thrilled to see how much room there is compared to the somewhat cramped crew cabin of the *Jupiter*. The space hotel module is virtually empty except for a small compartment with control panels near the upper end.

You push off with your hands and feet from one side of the module to fly to the other side 4 meters (13 feet) away. Soon you learn how to move your body in mid-air by swinging your arms and legs and manage to do slow-motion rolls and flips. It is an exhilarating feeling of freedom, like you have just learned to fly! You imagine you are Superman, flying from the bottom to the top of the module with your arms stretched out in front of you, or Spiderman, jumping from wall to wall.

Running over the cylindrical inside surface of the module, perpendicular to its long axis, creates sufficient centrifugal force to keep you pressed against the surface. You run a couple of rounds, then jump up and continue with a couple of somersaults.

You decide to try a trick you saw in one of the movies in the space camp: you push with an arm against the wall to start you spinning around your axis, then you extend and retract your arms to slow down or accelerate. It's really fun, but soon it makes you dizzy. After a rest, you decide to have a look at the Earth again.

Astronauts have always wanted more and bigger windows, but here in the space hotel any Earth gazer can be happy as there are four huge windowed domes, one for each of you. They are copies of the "Cupola" observation domes that were on the International Space Station, with six flat windows at the sides and a round one on top to ensure a superb view. When not in use, the windows are protected by shielding panels against micrometeorites that can scratch and crack the glass, but now these panels have been opened to reveal a stunning view of Earth.

The station is orbiting the planet in a vertical position, with the main docking hatch down. Since the windows are on the sides of the large module, the Earth can always be seen through each Cupola.

At the night-side of the planet, while crossing the southern border of Canada, a fellow crew member draws your attention to a magnificent phenomenon. He points at a couple of huge, circular waves of hazy green light that seem to dance over the Earth's surface, constantly whirling and changing shape. You realize that it must be the Aurora Borealis, also know as the Northern Lights. The Sun is currently in a very active state and sends out huge amounts of charged particles. The Earth's magnetic field guides them toward the polar areas of the planet, where the particles interact with the atmosphere and lose their energy in a staggering display of light. Normally it appears only further north, so you are lucky to be able to see it so far south. From directly under it the Aurora must appear to be covering almost the entire sky, but from space it looks more like a distant group of small, shimmering ghosts.

The copilot enters and reminds you that you should take the opportunity to play around some more, since time is going fast. He brings some toys to increase the fun, such as some small gyroscopes and a ball. You feel like kids on a playground as you bounce the ball against the walls and to each other, using different parts of your body. When jumping to intercept the ball, you have to estimate its trajectory quite accurately to catch it. While flying through the air, there is nothing you can do to change your velocity or your direction. An additional factor that makes it difficult is that the ball follows absolutely straight lines, except when you spin it for effect. Just as on Earth, the resulting air pressure differences on the sides of the ball will then make it follow a curved trajectory.

The small gyroscopes are fun too, as they stubbornly maintain their attitude when you push them around.

The pilot and copilot are constantly around to capture everything that happens on video. After the flight, it will be edited by Orbital Destinations, after which they will give you a copy as a souvenir of the flight.

The copilot puts some music on the sound system of the space hotel, and invites you to do some zero-g dancing. Soon everyone is rocking about, trying all those fancy moves you see in music video clips. You stop for a second when you realize how amazing this is: you are floating in space, looking at the Earth 250 kilometers (155 miles) below while you circle it at 7.5 kilometers per second (4.7 miles per second) and listening to some music. Feeling as if you are in heaven, you close your eyes to enjoy the moment.

Suddenly you see some streaks of light with your eyes closed! Somewhat upset you ask the copilot if something's wrong with you. He reassures you that this sometimes happens, as highly energetic atomic particles from space fly through your body and hit molecules in your eyes, causing swift emissions of light. It is nothing to worry about, although it reminds you that there is serious radiation out here, too much of which could be really harmful to your health.

GOING OUTSIDE

All too soon it is time to leave the "Freedom Flyer" and return to the spaceplane. Back inside you find the pilot is putting on a spacesuit similar to the one you have seen in the Neutral Buoyancy Facility at the space camp. He has been breathing pure oxygen through his launch helmet for the last half-hour, to expel all nitrogen from his blood. This is necessary to prevent him from getting "the bends" – the painful formation of nitrogen bubbles in the bloodstream due to the lower pressure inside the spacesuit. It is a problem first encountered by divers, when they came up too quickly to the surface from the high-pressure depths of the sea. Caisson disease, as the phenomenon is officially called, causes cramping and severe pain in the joints and possible paralysis or even death.

The air pressure inside the suit has to be lower than the normal pressure inside the spaceplane otherwise the suit would be too hard and inflexible to work in.

The pilot is going to put a small scientific instrument on the outside of the space hotel, where it will be measuring the effects of vacuum and radiation on a new type of plastic. Companies and institutes can rent space on one of the experiment racks on the "Freedom Flyer." Since the main income of Orbital Destinations is space tourism and support to scientific experiments is only a small additional activity, putting experiments on the space hotel is very cheap. Furthermore, the frequent visits to the hotel by the company's spaceplanes ensures very flexible launch and retrieval dates.

The copilot rubs an anti-fog mixture on the inside of the visor and puts the helmet over the head of the suited pilot. Next, the air pressure inside the suit is increased some 30 kilopascals above the cabin pressure. The difference makes it possible to detect any leaks by monitoring the pressure inside the suit: any pressure drop means that the suit is not airtight. Small leaks are normal and acceptable, as long as they do not cause the pressure to go down by more than 1.4 kilopascals per minute.

Now fully suited, the pilot floats inside the space hotel, toward the second docking adapter on the other side. This adapter doubles as an airlock through which an astronaut can get outside. After the pilot has entered the small cylindrical room, he closes the door behind him, pumps out all air and opens the outside door.

A spacesuit is in fact a miniature spacecraft, providing pressure, oxygen, water, cooling and protection to the person inside. Under the many layers of the suit itself, which protects against vacuum, extremely high and low temperatures, radiation and micrometeorites, an astronaut wears what looks like a pair of long underwear with a series of tubes running throughout it. The tubes circulate cooled water to keep the spacewalker from overheating inside the suit.

Under the actual helmet a so-called Snoopy Cap is worn. This is a fabric cap that contains earphones and a microphone for communication with the spaceplane and ground control. (They once were colored black and white and resembled the head of the cartoon character Snoopy.)

Through a window you watch the pilot come out of the airlock, holding on to special rails and secured to the spacecraft by a cable. The thought that there is nothing between him and the hostile vacuum except some thin layers of fabric and plastic makes you shiver. The side of the suit facing the Sun is heated to temperatures as high as 120 degrees Celsius (250 degrees Fahrenheit), while the shadow side may get as cold as minus 160 degrees Celcius (minus 260 degrees

Fahrenheit). The pilot does not seem to mind, and with gentle moves attesting to his spacewalking experience he carries on with the job at hand. The new experiment box is quickly secured to the rack, but the pilot stays out for a few minutes longer to enjoy the truly extraordinary view from the top of the space hotel. Looking down, he sees the sleek shape of the docked *Jupiter* outlined against the bright Earth, while he and the spacecraft are slowly turning around the planet. It is one of the few privileges in low Earth orbit still reserved for professional astronauts only. Due to the risks and complexities involved in a spacewalk, it is considered to be too dangerous for a space tourist.

When the pilot reenters, he brings with him a faint odor somewhat similar to that of burned metal. It must be the effect of the high temperatures on the spacecraft and his suit. It is the smell of space.

Soon the clock says it is evening again. Tonight, some crew members will be able to hook up their sleeping bags inside the space hotel to create some more room for those remaining inside the spaceplane's cabin. Obviously, the places near the large "Cupola" windows are the most favored.

8

SPACE STATIONS: GIANT CANS AND WHEELS IN THE SKY

HUMANS have gained experience living in space for extended periods since 1971, when the Soviets launched the first ever space station, Salyut. Space stations permitted cosmonauts to stay longer in orbit than before and provided much more room for experiments.

The Soviet space station program didn't have a very good start, however; the first crew sent to Salyut was not able dock because of a mechanical failure, while the second crew stayed on board for 22 days but died during their return. Due to the premature opening of a small ventilation valve during descent, all air leaked out of the capsule. Since the three crew members were wearing only overalls, they had no chance of survival. The Soyuz capsule landed automatically, with three dead bodies on board.

To prevent a repeat of this disaster, later Soyuz cabins flew only two crew members wearing complete spacesuits. Since suited cosmonauts required much more room, the safety measure could only be implemented at the expense of the third seat. Only when the new Soyuz-T was introduced in 1979, could three cosmonauts in spacesuits be housed within one capsule.

The second Salyut station experienced a major problem with its orbit correction system and was never crewed, but Salyut-3 was a success. It was a military station with an all-military crew, and was probably equipped with a large spy camera. Salyut-4 was a civilian system, but Salyut-5 was another military project.

The civilian Salyut-6 had an important new design feature: two docking units, making it possible to have a Soyuz docked to one side while the other unit could be used for Progress uncrewed cargo spacecraft. The Progress ships could bring fresh supplies and new equipment during a crewed mission, thereby making it possible to have longer flights on board the Salyut. It could also boost the station into a higher orbit with the use of its rocket engine, giving the Salyut-6 a longer lifetime than its predecessors; earlier Salyuts burned up in the atmosphere relatively soon when their orbits became too low because of the air drag in low Earth orbit.

The second docking unit proved to be of great value when Soyuz-25 banged into the main docking port but failed to connect with the station. The front docking unit was assumed to be damaged, but luckily the next Soyuz could still dock with the second unit at the back of the station. A spacewalk by the new crew to inspect the main docking unit proved that it was still in good condition and could be reused for later missions.

Between 1977 and 1981, 16 crews worked on board the Salyut-6. In total, the station was crewed for 676 days, during which unique astrophysical, geophysical, biomedical and material research was performed. The Russians learned much about life in space and the long-term effects of microgravity on the human body.

Salyut-7 was launched in 1982. Its 100 cubic meters (1,080 cubic feet) of pressurized volume housed everything to support long missions: a refrigerator, a television monitor (to enable the crew to see who they were talking to on Earth, such as their families), comfortable beds, a foldable shower, a treadmill and exercise bike, and a toilet with a nice wooden seat. There was room for a collection of recreational books and videotapes, and even musical instruments found their way to the station on board the Progress cargo ships.

The Salyut-7 station was crewed for a total of 34 months, with exceptionally long missions of 211, 150 and 237 days.

As the two docking ports also made it possible to have two Soyuz spacecraft docked to the station at the same time, long-duration crews could receive visitors. They usually stayed for a week. Many of them were guest cosmonauts from other Soviet countries, but some visitors even came from France and India.

The long-term missions proved to be very valuable. For instance, observations of the Earth by the cosmonauts helped to raise awareness of environmental problems in the Soviet Union, enabled maps of the vast Russian territory to be improved, and showed the Russian fishing fleet where large amounts of plankton, and therefore fish, could be found.

NASA launched its first space station, Skylab, in May 1973. It was basically a converted third stage from a Saturn V moonrocket, with its hydrogen tank now used as a two-level work and living space and its smaller oxygen tank used as a rather large garbage can.

The huge station was put in orbit on a modified Saturn V, but the launch was far from flawless: during ascent, a large piece of thermal isolation and one unfoldable solar array were torn from the station. Once Skylab was in orbit, the second solar array wing refused to open.

Only a few small solar arrays were left to provide power to the station, while the damaged isolation caused the temperatures in the station to rise to dangerous levels.

The first three crew members were quickly retrained to become space repairmen. They unfolded a special gold-sprayed parasol above the station to protect it from the Sun and lower the temperature inside, and also managed to free the blocked solar array.

Skylab had a huge inner volume of 347 cubic meters (3,740 cubic feet) available for its crew. The habitable space had an upper level that was used to house scientific equipment, food and other storages. There was still so much room left that the astronauts could do somersaults and run along the wall of the station.

The station's lower level contained three crew cabins (one for each astronaut), a galley, equipment for biomedical experiments and a toilet. For the astronauts it was a real luxury facility, as, for the first time, hot meals could be prepared. The toilet was an enormous improvement over the plastic bags used until then. Furthermore, there was a real shower that the crew was allowed to use once a week. On the downside, there was only one window, and even that had only been added on the insistence of the astronauts.

A separate section contained the airlock through which suited astronauts could go outside to exchange films in the large solar telescope mounted to the side of the station. The section also contained consoles to operate the solar telescope as well as cameras used to observe Earth. The docking facilities attached to this section allowed two Apollo capsules to be docked at the same time in case of an emergency.

Only three crews lived on board Skylab; the first for 28 days, the second for 56 days and the third for 84 days. The tight NASA budget did not allow further operation of the station, and it was already de-orbited in July 1974. NASA didn't launch another space station until the first element of the International Space Station was put into orbit in 1998.

In 1986, Russia launched Mir, a new generation space station. The main novelties were its six docking ports, to which additional modules were docked over the years. The station expanded from a single core module to a large complex that could house three permanent crew members for long-duration missions (see Figure 40).

FIGURE 40   *The Russian Mir outpost was the first space station that was continuously crewed. Like the International Space Station, it was built up from individually launched modules. [Photo: NASA]*

FIGURE 41 *The International Space Station, build by the USA, Russia, Europe, Japan and Canada, is still under construction. When finished, it should look like this. Nevertheless, the first space tourists have already visited the station. [Photo: ESA]*

FIGURE 42 *Inside, the living module of the International Space Station somewhat resembles a trailer home. [Photo: NASA]*

A new duration record was set when cosmonaut Valeri Polyakov stayed 438 days on board Mir. Astronauts from all over the world visited the station, and were launched with the new Soyuz-TM or with the American

143

FIGURE 43 *Even though the International Space Station is relatively new and modern, it is still rather cramped inside. [Photo: ESA]*

Space Shuttle. Five American and 11 European astronauts (one even went twice) stayed on board Mir for over 6 months at a time, gaining valuable experience for the International Space Station.

During its 15-year lifetime, the Mir station survived a collision with a Progress cargo ship, an internal fire, many power failures, leaking cooling systems and the fall of the Soviet Union. It was a true space station, where international crews learned to really live and work in space.

The private enterprise MirCorp owned Mir for some time before it was taken out of service; it had plans to convert the station into a commercial space laboratory annex space tourist hotel but could not find enough customers to keep the station running.

Mir's eventful life ended in March 2001, when it was safely de-orbited because of its old age and because Russia could not afford to both operate the Mir and fulfill its commitments to the International Space Station project at the same time.

In the mid–1990s, America, Canada, Europe, Japan and Russia started the International Space Station (ISS) project. It is a huge station that is still under construction, requiring a total of no less than 33 assembly flights by Space Shuttles and uncrewed Proton rockets, bringing modules and equipment (see Figure 41).

In its final form, the ISS is planned to have a mass of 450 metric tons

(992,000 pounds) and 1,200 cubic meters (12,900 cubic feet) of internal volume. This means that the crew will have about three times more room than was available on board the Mir station.

The ISS, although not finished, has been permanently occupied since November 2000. Although it was not originally planned, the station became the first space hotel when space tourists Dennis Tito and Mark Shuttleworth visited the complex. Space Adventures and MirCorp are both in the process of organizing more such tourist trips to the ISS with the help of the Russians (see Figures 42 and 43).

HOTELS IN SPACE?

Are there any real space hotels on anyone's drawingboard? Yes, MirCorp is seriously planning to launch a small space hotel to house its space tourists. The MirCorp orbital facility, currently named Mini Station 1, is planned to accommodate three visitors for stays of up to 20 days at a time. It is to have a lifetime of more than 15 years and will be serviced by both Soyuz crewed transports and uncrewed Progress cargo spacecraft.

The Mini Station 1 concept resembles a Soyuz with enlarged solar arrays and a large radiator wing, on which the return capsule is replaced by a habitation module. The small station would have two docking units (see Figure 44).

On a typical flight, the Soyuz would go first to Mini Station 1, where it would be docked for the two-week commercial mission. It would then fly to the ISS, where the Soyuz crew would transfer to the older Soyuz already docked to the station. The crew would return in this Soyuz, leaving the newer spacecraft as a fresh emergency escape vehicle for the ISS.

No other new space stations are planned for the near future, but there are plenty of concepts and ideas.

The Space Island Group has proposed building a huge orbital complex out of Space Shuttle External Tanks, the large brown cylinders between the Shuttle Orbiter and the two Solid Rocket Boosters. Normally these huge tanks are discarded just before the Orbiter starts to circle the Earth, but with just a small additional push the tanks could also be placed in orbit.

Tanks put in orbit during each Space Shuttle mission could be connected to each other to create a large wheel-shaped station that could provide a centrifugal force equal to half the Earth's gravity. Inside the outer

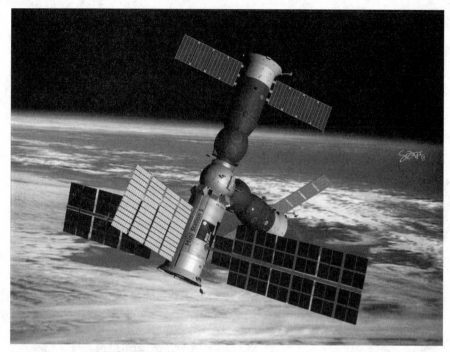

FIGURE 44   *The Mini Station 1 proposed by MirCorp. It would be specifically designed and put in orbit to act as a small space hotel for space tourists launched on board Soyuz capsules. [Photo: MirCorp]*

ring, space tourists and professional astronauts could live in a similar way as on Earth, being less vulnerable to space sickness, muscle atrophy and bone loss. Moreover, all kinds of medical treatments, such as the filling of a tooth, would be much easier to apply in partial gravity. In microgravity blood would float, obscuring the doctor's view and creating messy situations.

In the center spoke of the station the rotation would not create a lot of gravity because of its small radius. In this microgravity environment scientific research could be performed and space tourists could enjoy the feeling of weightlessness.

Similar space hotels have been studied by the Japanese Shimizu Corporation and the German aerospace company DASA. Shimizu's concept is a 64-module, 141-meter (463-foot) diameter wheel that can house 64 tourists (each guest has his/her own habitation module). The station would slowly rotate to create a gravity force equal to 70 percent of that on Earth (i.e. 0.7 g). Tourists would find it easy to walk around, especially people who would be less mobile on Earth, while the relatively slow rotation of about three rounds per minute would not cause dizziness or put intolerably severe loads on the structure. For promotion purposes,

Shimizu even made a short cartoon movie about a group of tourists visiting their space hotel.

DASA's designs are also wheel-shaped but somewhat larger, housing up to 240 people. Structural loads are minimized by a rotation of less than two rounds per minute, creating a 0.3-g gravity force in the outer ring of the station.

Checking in at a 2040+ space hotel

Large orbital space hotels will probably resemble the space stations envisioned by the Space Island Group, the Shimizu Corporation and DASA. Basic wheel-shaped stations were already designed in the 1950s by people such as Wernher von Braun, and Stanley Kubrick and Arthur C. Clarke put a huge one in their movie *2001: A Space Odyssey*. However, ring-shaped space hotels only make sense if we make them very large, and therefore they can only be built when mass space tourism has become firmly established and generates a lot of revenue.

Nevertheless, nothing stops us from imagining a tour through a hypothetical, 2040+ space hotel today!

Transport from Earth to the hotel will probably be with a fourth-generation spaceplane, based on airbreathing rocket engines and airliner-like flight and maintenance procedures (rather like the "Orion" spaceplane that can also be seen in the "2001" movie).

The vehicle will dock at the center of the wheel-shaped hotel, where a small section is de-spun with respect to the rotation of the space station. This makes docking much easier and is therefore also much safer than trying to connect with a rotating docking port.

Once the hatches are opened you can simply float from the spaceplane into the stationary part of the space hotel. As neither rotates, there will still be microgravity conditions in both.

Entering the spinning part of the station's center will also be simple, as the station rotates only about twice in a minute. This slow revolution results in a high rotational velocity in the ring, creating sufficient artificial gravity for walking. However, in the center of the wheel the rotation speed is low due to the small radius. The transition from the stationary docking module to the spinning axis of the hotel is thus very easy.

You will have to pull yourself in using the handrails to get from the stationary part into the slowly rotating part, and then further along to one of the tunnels that form the spokes of the giant wheel.

Moving outwards through such a tunnel, the radius and thus rotational velocity and artificial gravity increase. For a smooth transition, and to prevent accidents, elevators can be used to go from the hotel's center to the outer wheel. The hotel's main living areas will be located there.

Inside the main ring you will be able to stand normally in almost Earth-level gravity conditions. Like water staying inside a bucket spun around on a length of rope, you will be glued to the inner wall of the wheel, with your head in the direction of the station's axis.

Walking slowly will be just like on Earth, or maybe even easier if the artificial gravity is a little less than that on Earth (for instance, to help elderly and disabled people enjoy their holiday more). Running will, however, result in some peculiar phenomena. Sprinting in the same direction as the wheel is spinning effectively increases your rotational velocity, resulting in a higher level of artificial gravity. Running in the other direction, against the rotation, results in a lower rotation speed and thus less gravity. Running through the ring in one direction will make you feel heavier, while going the other way will make you feel lighter.

On a station where the artificial gravity is 90 percent of that on Earth, you could feel about 15 percent heavier or lighter when you run through the ring. In theory, using a motorbike you may get fast enough to fully compensate the station's rotation, at which point you would effectively stand still and thus start to float in microgravity!

Part of the check-in procedures will be a review of all safety and security aspects. In emergencies such as fires and meteor impacts, segments of the station can be evacuated and closed off. (Normally staff should handle all procedures, but the guests will be shown how to operate the emergency hatches just in case.) Fire extinguishers and oxygen tanks for breathing in heavy smoke will be distributed throughout the station.

In major emergencies, the whole station can quickly be evacuated using either a docked Orbiter or a series of permanently attached escape capsules. The Orbiters at the station's center can be reached via any of the tunnels in the spokes. The escape capsules are distributed over the outer part of the ring of the hotel, where the rotation ensures that the lifeboats fly away when detaching.

A tour through the space hotel will take new guests to the different parts and facilities of the station. On the first space stations, and probably on the first space hotels, living, eating and sleeping will have to be done in microgravity conditions. The large space wheels of the future will, however, have both gravity and microgravity areas.

The ring of our hypothetical space hotel will house all the places where it is better and easier to have gravity. These will include typical hotel accommodations such as bars, fitness rooms, medical facilities, etc.

A large module in the ring is devoted to environment control and life support. It is here that the atmosphere is filtered, carbon dioxide is removed and oxygen is replenished. The system will probably be self-sufficient to a high degree, requiring only a minimum of fresh water and oxygen to compensate for losses. A mixture of chemical and biological technology will clean waste water, break down solid waste products and generate oxygen from carbon dioxide.

The hotel's restaurant will provide specially selected food, minimizing waste to ease the burden on the environment control and life-support systems.

The guest rooms will probably be a lot smaller than you are used to on Earth, since volume will be at a premium on board any space station. The furniture of the rooms may be like that in a small camper, with foldable beds and tables to save space.

Each cabin will probably have a window, although the rotation of the station will make it difficult to keep the Earth in view at all times. A kind of periscope system with rotating mirrors could help in this respect. There may also be different classes of rooms, with the largest and most expensive suites incorporating extra-large glass domes for an unprecedented view.

However, the hotel's main attractions will be located in the hub of the wheel, where microgravity conditions are possible. There we could find microgravity swimming pools, other novel weightless sport facilities and free-float lounges where guests can have a magnificent view of Earth while enjoying a drink. The microgravity area may also house special suites for those guests who want to go for a full living-in-weightlessness experience.

In the center of the hotel, a de-spun observatory with powerful telescopes could provide breathtakingly clear views of the stars and planets. Guests could also zoom in at parts of the Earth and look for their own house. Furthermore, there may be laboratories where guests could perform their own microgravity experiments with various fluids and plants, serious or just for fun and education.

In the next section, we'll take a more in-depth look at the novel entertainment possibilities that large space hotels could offer.

MICROGRAVITY SPORTS AND RECREATION

Can you do sports in space? Sure, but until now there as been neither sufficient room nor time for organized sports on board spacecraft. However, once sufficient volume on board future space stations becomes available for microgravity exercises, people will want to play games. What sports will entertain space tourists of the future?

Traditional terrestrial sports would have to be played quite differently as many two-dimensional games like football or tennis would turn into complex three-dimensional activities. Moreover, players would be floating and not always be in contact with the ground, walls or ceiling. In mid-air, out of reach of any fixed surface, it would be difficult to change direction.

A new form of soccer or rugby may have to be created, with players pushing each other off in different directions in mid-air, or using heavy floating objects to push away from when no fixed surfaces or objects are close enough. Players could be launched by team members who have their feet attached to a wall, to intercept opponents and push them away from the ball.

A goalkeeper would need to be attached to his goal with bungee cords, otherwise he would float away after jumping for the ball. Goalkeepers may be selected from particularly heavy people, so that their high inertia would prevent them from being propelled into the net after catching a ball.

A two-dimensional soccer pitch with lines to indicate different areas is sufficient for play on Earth, but in space everything would have to be scaled up by one dimension: the field becomes a three-dimensional space and the lines would have to be translated into surfaces dividing the playing volume into different zones.

One thing that would definitely need to be changed with respect to terrestrial soccer is: no spitting!

Apart from adapted versions of existing sports, whole new forms could be created. People could actually fly, pushing themselves against the air with large foil wings using small battery-powered ducted-fan engines to move around. Indoor aerobatics could become a really exciting but safe sport!

And what about a cylindrical or spherical room lined with trampolines that would allow tourists to jump and float from one side to another? You could push off from the wall and try to land on your feet on the other side, perhaps doing a few rolls in between (see Figure 45).

Patrick Collins – an economist and well-known advocate of space tourism, working for Tokyo University – together with several employees of the Japanese construction firm Hazama Corporation, designed a microgravity gymnasium and a space stadium. Both could be attached to a space station or space hotel.

Slow-spinning stadiums for low-gravity sports may also be interesting, allowing you to jump further and higher than is possible on Earth, with

FIGURE 45 A main attraction of a future space hotel would be a microgravity sports center, where people could freely enjoy weightlessness. [Photo: Space Island Group]

less danger of hurting yourself. Athletes may, for instance, use a $\frac{1}{6}$-g facility to train for the first Olympic Games on the Moon!

Ball games would become more complex in such a place, as the balls would follow strangely curved trajectories owing to the rotation of the playing field.

The possibilities of large amounts of water in microgravity conditions are also intriguing. Patrick Collins, for instance, imagines space tourists playing with large spheres of water. You could get such a sphere spinning like a pancake, pop the middle and go through it, which should be good for a couple of hours of fun.

Swimming in a large ball of water in a microgravity pool should be great too, but because of the lack of neutral buoyancy you would not automatically float back to the surface after a dive. It would take some effort to get out of the floating ball of water, so mini scuba equipment may be necessary for safety reasons.

The next step in aquatic space entertainment would be a large partial gravity swimming pool, another large orbital facility envisioned by Collins and the Hazama Corporation. Their design involves a large spinning cylinder of some 10 meters (33 feet) long and 20 meters (66 feet) across. The rotation would force the water against the inner wall, where it would collect inside a trough ringing the cylinder. An area of the wall would remain dry to function as a kind of beach.

A gravity force much lower than on Earth would mean you could jump high out of the water and even traverse all the way to the other side of the wheel. Or you could throw armfuls of water into the air, where it would form large floating, wobbling bubbles. Diving boards could be very high, with jumpers gently floating into the water rather than diving down fast.

Other forms of recreation would benefit from a whole range of new microgravity possibilities: dancers could perform ballets that are unthinkable on Earth; acrobats in a microgravity circus would invent new, exciting performances; sculptors could create the most bizarre and daring works of art without having to worry too much about the stability of their designs; fashion designers would find new inspiration in the possibilities for microgravity clothing and spectacular hair styling. And what about filming realistic space movies on location? Instead of in simulations and models, or in planes simulating microgravity by following parabolic trajectories (as for the filming of *Apollo 13*), actors could actually be in space.

SPACE LOVING

Do astronauts have sex in space? It is a question that keeps coming back, as many people are intrigued by the idea of romance in microgravity.

NASA's regulations do not prohibit any sexual activity among their astronauts in space, but the agency has always vigorously denied that any such thing has ever happened during one of their flights. The Russians have also never reported anything about their cosmonauts getting a bit too interactive. There is a rumor that the Soviets planned to send a married couple in space for sex experiments in the mid-1960s, but such a flight was never carried out. In any case it is difficult to imagine that the Soviet Union would waste time with such research while it was frantically trying to win the Moon race.

Nevertheless, some journalists insist that sexual research has taken place in orbit. An alleged secret NASA document posted on the Internet even describes experiments involving a male and a female astronaut. It describes the lovers trying various positions with the help of elastic belts and inflatable tubes to hold them together. (The report noted that "the investigators explored a number of possible approaches to continued marital relations in the zero-G orbital environment ...".) The experiments are claimed to have taken place during Space Shuttle mission STS-75, which had not even been launched when the report was on the Internet. In fact, when the STS-75 eventually flew, it had an all-male crew. Some journalists, probably looking for a way to spice up their stories, even now insist that this spoof report is real.

The fact remains, though, that mixed gender crews have been flying in both the Space Shuttle and on board the Russian Mir station. One Space Shuttle crew even included a married couple. However, the lack of privacy on board the cramped spacecraft and the rather tight schedules have very probably prevented any physical romantic activities. Moreover, missions involving mixed gender crews have been relatively short, short enough for most people to keep their sex drives under control. (The combination male–female does not have to be a prerequisite for any sexual activity, of course.)

On long-duration missions, such as a flight to Mars, the situation could very well be different. Experience on crew interactions on, for instance, Antarctica expeditions, with groups of men and women living in isolation inside polar bases for months at a time, shows that social interactions are of major importance. Romantic feelings, and with it emotions such as jealousy, obsession and guilt, can make or break crew morale and have an enormous influence on the success of a mission.

On first thought, Earth orbit appears to be the ultimate place for some romance: always a starry sky, 16 sunsets a day and a beautiful view. Moreover, floating intercourse means no more squashed legs and arms. "Orbital lovers" may, however, experience some "rendezvous and docking" problems, requiring special aids to stay together. The use of special four-legged shorts has been suggested. An American Sunday school teacher has even patented a system of straps and loops to allow one partner to exercise control of the movements of the hips of the other partner during love-making. Others propose special furniture and fittings to help people to stay together.

Former NASA consultant G. Harry Stine even suggests that assistance of a third astronaut holding one of the copulating couple in place would make things much easier. In his book *Living in Space*, he claims that the neutral buoyancy tank used for astronaut training at NASA's Marshall Space Flight Center has already been used as a proving ground for this technique. People taking part in these very unofficial investigations supposedly call themselves the "Three Dolphins Club", since two mating dolphins sometimes also get help from a third.

It is not difficult to imagine that space tourists will be very interested in exploring the possibilities of making love in microgravity. Honeymoons in orbit may very well become the best-selling holiday trips in the future, but not before large enough space hotels are built to offer sufficient privacy. Science fiction author Allen Steele even describes a "Fornicatorium" on board a tourist space cruiser, specifically intended for some intimate, undisturbed 0-g fun.

SPACE FOOD

In space, the type and amount of food the human body needs do not change much; astronauts in orbit require about 2,700 calories each day, similar to what they need on Earth.

Eating is not only of physical importance but also has a strong psychological effect because meals act as natural breaks during a busy day, when all crew members leave their individual activities for a while to eat together. A good meal is a wonderful moral boost on long-duration missions.

Ideally, astronaut meals should therefore resemble Earthly breakfasts, lunches and dinners as much as possible. However, the microgravity conditions in Earth orbit mean that the type of food and the way it is eaten are quite different to what we are used to on Earth.

First of all, the lack of gravity means that all food must be prevented from floating away; airborne crumbs could get stuck in equipment and cause malfunctions or lodge themselves in lungs and eyes, while free-flying blobs of liquid may cause short-circuits. Tortillas are therefore favored over ordinary bread because they don't crumble, and liquids are contained in squeeze packs with drinking straws. Sticky sauces like mayonnaise and ketchup are ideal for use in space.

Second, to save mass and storage space and to be able to keep it without cooling, most astronaut food is vacuum packed and dehydrated. For longer storage, some foods are also heat sterilized to prevent bacteria and moulds from making them inedible. They are sealed in conventional cans or plastic pouches.

Water, a by-product from the electricity generating fuel cells on board the Orbiter, can be added to dried vegetables, fruit, egg powder and pre-cooked pasta from a water dispenser to produce fast and easy meals. Hot soup, for instance, is obtained by injecting hot water in a plastic bag filled with soup powder. You knead the bag and attach a straw, then squeeze the soup directly into your mouth.

Unfortunately, not all drinks can be treated in this way: adding water to dehydrated orange juice only results in a collection of floating grains, and the powder of whole milk creates an unappetizing lumpy substance. Fruit juice is therefore synthetic, like lemonade, while only low-fat milk powder is available for breakfast (see Figures 46 and 47). However, normal dry food like small crackers, nuts and candies, are already suitable for use in space.

Not only is the food in space different from that on Earth, but your taste also changes. The increase in liquids in your head (as described in the previous chapter) congests the sinuses, and because taste not only depends on the tongue but also on the nose, the flavor of food will diminish. It is similar to what happens when you have a cold. The effect is felt almost immediately after reaching orbit, but diminishes after a few days when the amount of liquid in your body finds a new equilibrium. During a three-day trip, you may no longer fancy what is normally your favorite food, while you develop an appetite for strong-tasting dishes you would never put on your plate on Earth.

During their first days in space, astronauts often prefer spicy food; salt and pepper are essential for a space meal. You can't use normal salt or pepper since the grains would float away, but they are available as liquids that stick to the food.

In the International Space Station, complete meal packs are available for breakfast, lunch and dinner. They contain bags of dehydrated foodstuffs to be injected with warm water and cans that can easily be heated in a small electric oven. On the Shuttle Orbiter, one crew member can prepare meals for four people in about five minutes (talk about fast food!).

The different food containers are put on trays, one for each astronaut, with the use of Velcro or magnets. The trays can be attached to the legs or any other surface with adhesive straps; tables and chairs are not required (see Figures 48 and 49).

Normal forks and spoons can be used on solid food, if they are first immersed in tasteless gelatin or sauce to make the food stick, as long as you don't make any sudden starts, stops or spins. Knives are not required because the food is already cut in single-bite chunks.

For some food, like crackers and candies, it is much easier to simply let them float around and grab them out of the air with the mouth. Ever tried to catch peanuts in your mouth? In orbit this is a very simple trick.

Soup and beverages can also be allowed to float: liquids in microgravity assume a ball shape and can be caught in mid-air and sucked into the mouth. You must take care, of course, that the blobs are not lost and allowed to contaminate sensitive equipment.

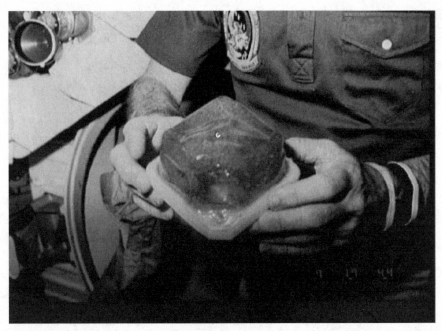

FIGURE 46   In space, orange lemonade is stored as powder in a bag. An astronaut has to inject it with water and knead the bag to make it drinkable. [Photo: NASA]

FIGURE 47   Cooking in microgravity is very difficult, therefore astronauts mostly eat heated food from cans and bags. [Photo: NASA]

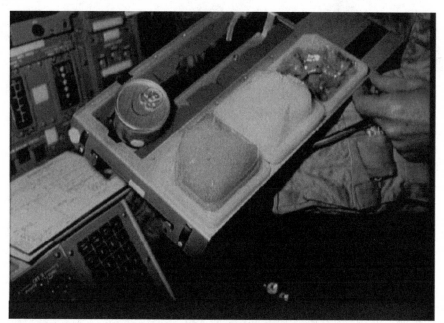

FIGURE 48 In microgravity, food items have to be attached to a tray to prevent them from floating away. On board the Shuttle Orbiter, the menu consists of canned food, reconstituted dried food and drinks made from powder. [Photo: NASA]

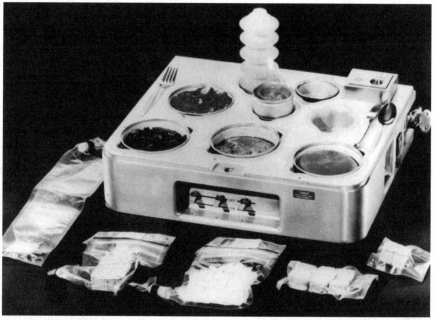

FIGURE 49 On board the Skylab space station, astronauts used this kind of food tray and heater in one. Future space tourists may desire a more appetizing way of eating. [Photo: NASA]

The Shuttle Orbiter menu includes over 70 food items and 20 beverages, ensuring different menus every day for six consecutive days. A typical day's menu includes orange drink, peaches, scrambled eggs, sausages, cocoa and a sweet roll for breakfast; cream of mushroom soup, ham and cheese sandwich, stewed tomatoes, bananas and cookies for lunch; and shrimp cocktail, beefsteak, broccoli au gratin, strawberries, pudding and cocoa for dinner. There is no refrigerator or freezer on board, but a special kind of substance that tastes like ice cream but does not require cooling has been developed especially for use in space. Small bags of this stuff are popular souvenirs sold at space museums and visitor centers.

Nobody has to do the dishes: all food is packaged in plastic that is simply thrown away as garbage. For relatively short-duration flights this is much more efficient than wasting water and precious time on cleaning.

Short-duration flights with tourists paying hefty ticket prices will probably include relatively tasty, appetizing meals, with more fresh food than is usual on Space Shuttle and Space Station missions. The careful diets of professional astronauts who stay in space for weeks or months, with balanced amounts of calcium, natrium, nitrogen and other minerals to help the body maintain healthy muscles and bones in orbit, will not be required for short tourism flights.

As on any luxury cruise or holiday, meals will be important for space tourists, and spaceline operators will have to take care that menus are available to suit the different tastes, cultures and religions of the passengers and that they are varied and appetizing.

Pizza Hut has already developed a vacuum-packed space pizza which can be kept for lengthy periods without freezing. It was eaten by Russian cosmonauts on board the International Space Station; footage of the event was later used in a TV commercial.

In Japan, Nissan Food Products, the maker of Cup Noodle, is collaborating with the Japanese Space Agency (JAXA) to develop instant space noodles for the astronauts on board the International Space Station.

Whether alcohol will be allowed on board spacecraft remains to be seen, as its effects on the human body in microgravity have not yet been studied. Too much alcohol and hangovers will not exactly help to overcome space sickness and will only add to orientation problems. Drunken people could also easily cause dangerous situations inside a spacecraft. Moreover, vomiting in microgravity can create quite a mess.

Cosmonauts and astronauts have drunk small sips of alcohol on board the Mir space station, usually to celebrate special occasions such as birthdays or the breaking of a long-duration space flight record. However, it is not likely that space will remain a teetotaler's heaven: students from the Technical University of Delft in The Netherlands have already developed a beer tap that can be used in microgravity. It was even tested during a parabolic flight campaign for students, sponsored by the European Space Agency.

The development of astronaut food has resulted in a lot of similar products being developed for a terrestrial market: instant, dehydrated foods and drinks are now sold in every supermarket, while space food studies have given us invaluable knowledge and experience in the preparation of special medical diets.

It is almost certain that future space tourist meals will not resemble the all-containing pills we were promised in the 1960s. At that time food scientists were looking for nutritional substitutes to feed astronauts on long-duration missions. The research suggested such unappetizing solutions as bite-sized cubes covered in edible gelatin, food puree to be squeezed out of a kind of toothpaste-tube and dry foodsticks.

The results, however, didn't go down very well with either the astronauts or the general public: "The gourmet's nightmare of a more distant future" appeared in the *Wall Street Journal* in 1966, while another headline concluded "Space Food Hideous But It Costs A Lot." Astronaut John Young agreed and smuggled a corned beef sandwich on board the five-hour Gemini 3 flight in 1965 to give it to his flight partner Virgil Grissom. The incident resulted in a Congressional investigation and the first ever official reprimand of an astronaut. Since then, space food improved and resulted in the meals now available on board the Shuttle Orbiter and the International Space Station.

Real cooking will remain very difficult until rotating space stations with artificial gravity are built. In such places special diets will not be needed and eating and cooking can be very similar to what we now do on Earth. Rather boring, isn't it?

SWEET DREAMS

Sleeping in space is strange: in low Earth orbit, days last only 90 minutes and there is no gravity to keep you and the sheets on a bed. The lack of gravity is handy for interior design, however: since there is no physical up or down, astronauts can sleep on the floor, against the walls and even on the ceiling.

The Shuttle Orbiter and modern space stations are equipped with small private rooms about the size of a phone booth where astronauts can

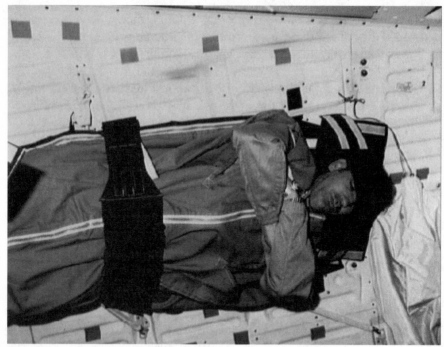

FIGURE 50 *Strapped down, you can sleep against a wall or the ceiling in microgravity. The band over the astronaut's face is to prevent head movements, which could cause neck problems. [Photo: NASA]*

snuggle up in their fireproof sleeping bags. The mattress they strap themselves against is little more than a plank covered with a soft material, since they are not actually lying on it anyway.

However, the Shuttle Orbiter only has four bunks, and the International Space Station also has limited accommodation. The astronauts who do not have a bunk attach their sleeping bags to their seats or in a quiet corner of the station and sleep without the benefit of a mattress (see Figure 50).

Dutch European Space Agency astronaut Wubbo Ockels invented and used a sleeping bag that put some pressure on the body: it had a rubber ring that could be inflated or deflated to vary the tension, mildly squeezing the astronaut inside to simulate the weight of bed covers.

Astronauts are scheduled to sleep eight hours each mission day. However, the lack of gravity and the 16 sunrises and sunsets per day do have an effect on the astronauts' sleep, as research on the sleep patterns of astronaut Jerry Linenger indicated.

During three two-week periods in the beginning, middle and end of his nearly five-month stay on board the Russian space station Mir in

1997, Linenger's sleep patterns were logged by NASA scientists. The study showed that after three months in orbit Linenger slept less soundly and for less time than he did on Earth. His internal clock appeared to get confused in space. "I lost my sense of day and night," Linenger said. "Every 45 minutes: light dark, light dark. So biorhythms are totally out of whack."

As a consequence of sleep deprivation, astronauts may be less alert, which may hamper performance on long-duration spaceflights. Linenger recalled that his fellow crew members sometimes fell asleep in the middle of a conversation.

Space sleeplessness is probably caused by changes in the brain's endogenous circadian pacemaker, a pinhead-size bundle of nerve cells that controls the body's sleep cycle. Light hitting the retina of the eye helps to determine the time of day, but in orbit the 90-minute days throw off this mechanism. In addition, the lack of gravity means that astronauts expend much less energy than on Earth and do not tire as easily.

Researchers offer a variety of remedies for sleeplessness in space, including sleeping pills, doses of a hormone believed to control the sleep cycle, and the use of bright lights to mimic sunlight. Physical exercise to counteract the effects of microgravity may also help. The trick appears to be to keep the body on a strong 24-hour cycle, forcing the biological clock to follow the same timetable as here on Earth.

However, for short-duration flights these problems are of no importance. Other than the excitement of the flight, nothing should keep space tourists from enjoying a good night's rest.

Waking up in space is probably just as difficult as it is on Earth, although Space Shuttle astronauts are helped a bit by the people of Mission Control, who get them out of bed by putting music on. They play a different song each day, usually a favorite of one of the astronauts and sometimes a tune specially requested by a family member. A couple of bands and singers even re-recorded songs with lyrics adapted to the Shuttle mission.

The tradition was kept when the famous NASA Sojourner rover landed on Mars. Each Mars morning, before the little wheeled robot began its investigation of the terrain, the ground control team played a fitting wake-up song such as "So Far Away" (Dire Straits) or "Love Me Like A Rock" (Paul Simon)!

Astronauts on board the International Space Station use a conventional alarm clock, fitting to their image of being more permanent inhabitants of space than the astronauts of the short-duration Shuttle missions.

PETS IN SPACE

Would you want to take an animal into space? For short trips it will surely be prohibited, but clients planning for long-duration stays in a large orbital hotel or Mooncruiser may want to take their pet with them.

You will remember that animals went into space before people did: in 1957 the Russians launched the dog Laika in orbit on board their second spacecraft, nearly three and a half years before Yuri Gagarin became the first man in space. America launched the chimpanzee Ham as a trailblazer for their first astronauts. Animals were also the first to fly to the Moon and back on board the Russian Zond test capsules.

Recently, the first Chinese to go into orbit were six mice and a bunch of fruit flies. They were inside the second Shenzhou test capsule the Chinese launched as a prototype for their first crewed spacecraft and were returned to Earth in a reentry capsule.

Most pets in orbit will experience the same space sickness symptoms as humans, and will also lose bone and muscle mass during longer flights. An additional problem is, of course, that it is difficult to explain to a rat what is happening to it in orbit, let alone convince it to follow a rigorous training schedule to maintain its strength.

Moreover, many animals are ill equipped to live in microgravity as they have no hands, claws or tails with which to grasp. Experiments have shown that rats, after some time of confusion and disorientation, soon learn to hold on to pipes and rails to keep from floating away. A dog will not be able to do that.

Fish born in microgravity also adapt to the strange conditions; they orient themselves to the light, whereby the brightest part of the spacecraft is "up". However, fish born on Earth never know how to determine "up" or "down" and are unable to function normally. Similarly, fish born in orbit brought down to Earth never learn to adapt to gravity.

An important question is whether animals would enjoy or could even be comfortable with life in orbit or inside an enclosed planetary base. Animals are not explorers and their idea of "fun" can be quite different from ours. Forcing them to live in space may be cruel.

On long-duration space missions, such as a trip to Mars, animals could, however, provide a boost for morale as well as act as a source of food.

For similar reasons, it may be nice to have some plants on board spacecraft. Unfortunately, plants also have some difficulty growing in microgravity. For instance, gravity determines which way plant roots grow; as there is no direction in microgravity, roots grow in all directions.

A solution to many of these problems is to put plants in a small centrifuge to create artificial gravity; relatively slow rotations are sufficient to direct the plant's growth. Neon tubes in the center of the centrifuges can provide the required light and make the upper part of the plant grow in the opposite direction to its roots.

Furthermore, bioengineering may result in plants that are better adapted to life in orbit.

Which, if any, animals and plants would we take into space? Would they fully adapt? Are we going to bioengineer space animals with extra arms and thin skeletons, perfectly adapted to live in microgravity but not able to live on Earth? Should we even try?

9

THE END OF THE TOUR

THE short, 30-second burst of the rocket engines pushes you back into your seat as the *Jupiter*'s maneuvering rocket engines fire into the direction of the vehicle's velocity. The thrust slows the plane down with some 300 kilometers per hour (190 miles per hour), which is sufficient to change its orbit from a circle into an ellipse that crosses the atmosphere. The aerodynamic drag brakes the vehicle further. You are now slowly falling back to Earth.

Just an hour ago you undocked from the Space Hotel. You watched the large, winged can slowly drift out of view. You feel sorry that your space adventure is almost over but look forward to the exciting, fiery reentry into the atmosphere.

In preparation for the landing, you have put on your suit, boots and helmet again and strapped yourself back into your seat. Everyone helped to ensure that no loose items were left lingering inside the cabin.

You drank about one and a half liters of juices and lemonade to compensate for the water you have lost in orbit due to the changes in your body's fluid balance. In addition, you swallowed eight salt tablets to help to keep the water inside. This will help to prevent problems when, back on Earth, gravity will once again pull fluids down in your lower body; if insufficient water remains in your head you may faint.

The pilot informs you over the intercom that weather at the spaceport is good, that permission for landing has been given by Air and Space Traffic Control on Earth and that he expects a smooth ride down. Secretly you have been hoping for bad weather, so that your trip would have been extended for some time.

With the old Shuttle Orbiter, landings were delayed many times because of the weather. For instance, the Orbiter would remain in orbit when there was any lightning or rain anywhere within 55 kilometers (34 miles) from the landing strip at the Kennedy Space Center. Sometimes it even had to land in the southern California desert because of the stringent weather constraints. However, the spaceplanes of today are not much more sensitive to rain, hail, lightning and low clouds than normal airliners.

For the next half hour after the de-orbit burn, not much seems to change: you are still in microgravity and the Earth's surface is still sweeping by under you. The pilots turn the spaceplane over 180 degrees, as it was flying backward when firing its rocket engines to break out of its circular orbit. Now you are flying forward again, with the vehicle's nose up some 30 degrees to align its heat-resistant belly with the wall of air it is just about to contact.

After a while you notice that the curvature of the horizon is flattening, and that the part of the Earth you can see through your window is getting smaller but slightly more detailed. You started from 250 kilometers (155 miles) high and now the altitude indication on your computer screen shows that you have dropped to about half of that.

The display also shows that the spaceplane's velocity is slowly dropping due to the increasing atmospheric drag. The deceleration meter indicates that there now is a tiny g-force acting on your body and you are no longer in microgravity.

At an altitude of about 120 kilometers (75 miles) you start to notice an orange-yellow glow shining through the window as the spaceplane is bashing into the thicker part of the atmosphere. The friction with the atmosphere heats up the air so much that the molecules are ionized and emit light to rid themselves of the excess energy. You can now also feel the rapidly increasing g-force as the rate of deceleration increases.

Soon the whole spaceplane is enveloped in a blazing, bright white-orange plasma. The ionized air interferes with the radio signals to and from the vehicle and makes direct communication impossible. In the past, this so-called "black out" phase was a nerve-racking experience for mission control on the ground, because there was no way to check what was happening to the returning

FIGURE 51 *Capsules and spaceplanes returning from space are heated to extreme temperatures due to the friction with the atmosphere. Without heat shields they would burn up. [Photo: ESA]*

spacecraft and its crew. The *Jupiter* can, however, send radio signals up through the wake of the ionization, to a communication satellite that relays the data to the ground.

The buffeting of the air outside makes the spaceplane shudder a bit. The pilots inform you that control over the vehicle is now slowly shifting from the small attitude control rockets to the conventional elevons and rudders on the wings and tail of the vehicle.

The nose and the edges of the wings reach a temperature of about 1,650 degrees Celsius (3,000 degrees Fahrenheit), higher than the melting point of steel (see Figure 51).

The *g*-force on your body increases further and you find it slightly uncomfortable; your body has already become a little more used to living in microgravity and, furthermore, you are not lying on your back as during the launch, but instead taking the deceleration force in a more vertical direction. Rubber pillows inside the pants of your flight suit automatically inflate to help to keep sufficient blood in your upper body by compressing your legs. Even though the forces you are subjected to are not very high, your body may have problems pumping enough blood to the brain after even a few days in microgravity. A "black-out" would not be a big problem but is not exactly enjoyable. The g-force reaches a maximum of one-and-a-half times your normal weight, then slowly decreases.

For almost 15 minutes all you have seen out of the window is the bright light emitted by the plasma. Then, suddenly, it disappears and you see the familiar bright blue of the upper atmosphere.

As you return to your normal Earth-weight the spaceplane breaks through a cloud layer to reveal the Florida landscape below.

The automatic pilot system gently steers the spaceplane through a series of wide-curved turns to further decrease the velocity. The vehicle is now a glider, flying without the help of any engines. Your approach is announced to the people at the runway awaiting your landing by a double sonic boom, caused by the shock waves from the nose and the wings of the spaceplane. At an altitude of 15 kilometers (9.5 miles) the spaceplane's velocity falls to subsonic. The vehicle is now approaching the landing site, following a computer generated "tunnel in the sky" that guides it to the runway by use of a satellite positioning system.

You are coming in about 6.5 times steeper than with a normal airliner, due to the relatively poor aerodynamic lift on the spaceplane. The spaceplane design is a compromise to fit contradicting technical demands for the different flight phases and is not optimized for subsonic flight. In just 5 minutes you have dropped from an altitude of 15 kilometers (9.5 miles) to ground level. The final approach is a bit scary; you know there is no second chance for a landing, as the vehicle lacks both the engines and the aerodynamic lift to abort and fly around for another try. However, you know that the *Jupiter's* automatic pilot is a very sophisticated piece of technology and will not let you down. And in any case, the pilots are well trained to take over from the computer, having practiced the landing hundreds of times in the simulator and with a specially equipped training aircraft.

Only moments before touchdown the landing gear is extended, so that the air turbulence caused by the open doors, undercarriage structure and the wheels does not disturb the aerodynamics of the spaceplane too soon.

A single, not too hard bump announces that the wheels under the wings are rolling over the runway, the nose wheel follows seconds later. Shortly after, a parachute is released from a canister just under the vertical tailfin of the spaceplane to help to brake the vehicle. As the *Jupiter* rolls to a standstill, the brakechute is released to prevent it from getting entangled with the spaceplane.

Like almost everything else on board, it will be reused for another flight.

As you sit in the quiet spaceplane, waiting for the ground crew to attach the mobile stairs, you feel extremely heavy. The gravity on Earth seems to crush you in your chair. Nevertheless, you have to suppress the urge to jump out and try to float to the ceiling; you know it will no longer be possible and, moreover, you have been warned that sudden movements can make you feel rather sick so soon after coming back from orbit. It will take some time for your vestibular system to readjust to permanent 1-g conditions.

The flight is over (see Figure 52). All that remains is the gala dinner this evening, during which you will receive your Astronaut Wings. But as the ground crew helps you out of your seat and you carefully stumble down the stairs to be greeted by family and friends, there is only one thing on your mind: "When can I go up again?"

FIGURE 52   In the 1960s, X-15 spaceplanes routinely flew to the edge of space and reached velocities of over six times the speed of sound, then glided back to Earth and landed like an airplane. In this picture, the bomber plane that airdropped the X-15 flies by the rocket plane that has just landed. [Photo: NASA]

10

RETURNING FROM SPACE

T HE first human being to experience a return from space to Earth was Yuri Gagarin. It was not a comfortable experience at all. First his landing capsule inadvertently remained connected to the orbital service module (the part of the spacecraft containing the batteries, tanks, rocket engines, electrical systems, etc.) by a bundle of electrical cables that failed to disconnect properly. Both parts of the spacecraft tumbled back to Earth together, wildly gyrating and threatening to bump into each other. Only after the superheated air of reentry burned through the wires was his spherical capsule released. It then naturally assumed its aerodynamic equilibrium attitude, with the reentry shield positioned correctly to protect Gagarin from the onslaught of the fiery atmosphere.

ROUGH LANDINGS

Gagarin's Vostok 1 capsule lacked the rocket system installed in later Russian spacecraft to cushion the touchdown. To avert a hard landing

Yuri was required to blast himself out with his ejection chair and finish the descent by parachute. The capsule also came down under a parachute. Yuri landed near the village of Smelkovka, not far from where he had taken flying lessons 13 years earlier.

For a long time, the Soviet Union claimed that Yuri actually landed inside the capsule. They were afraid the flight may otherwise not have complied with the rules of the FAI (Fédération Aéronautique Internationalationale, the International Aeronautical Federation), which require a pilot to land inside his vehicle for various aeronautical records to be recognized.

The first NASA astronauts had their share of troublesome returns: Gus Grissom, the second American to make a suborbital jump into space, almost drowned when the hatch of his capsule inadvertently blew open after a successful landing at sea. The spacecraft quickly filled with water, and Grissom had to swim for his life. As he had already removed his helmet and detached the suit's oxygen supply tubes, his suit began to fill up with water through the open inlet lines. The capsule sank but Grissom was rescued by helicopter just before the weight of the flooded suit could drag him down: the unfortunate astronaut had almost followed his waterlogged spacecraft. In 1999, however, the capsule was discovered in surprisingly good condition, and retrieved from the bottom of the ocean.

Astronaut John Glenn, the first American in orbit, also went through a stressful event when an onboard sensor indicated that his heat shield had detached. This vital layer of protection against the heat of reentry was designed to come loose, and thereby deploy a kind of airbag to cushion the landing in the ocean. However, this was not supposed to happen until well after the parachute was deployed. Post-flight analyses showed that the sensor's information had been false, but for some agonizing minutes Glenn didn't know whether or not he would burn up during reentry.

Although bringing back a 7.5 kilometers per second (4.7 miles per second) flying spacecraft from orbit down through the atmosphere to a soft landing was something relatively new in the early days of human space flight, surprisingly no serious accidents occurred. That is, until the disastrous flight of the first Soyuz in 1967. The new Russian spacecraft was intended to accommodate a crew of three, but for the test flight only cosmonaut Komarov was piloting the craft. Immediately after arriving in orbit, Komarov had problems when one of the solar panels failed to deploy. Although the spacecraft now only received half of the planned solar power, an attempt was made to maneuver the Soyuz. This failed because of problems with the small steering, thruster control system. The decision was then made to bring Komarov back. Reentry through the atmosphere went well and the drag chute, which was meant to slow the capsule down and pull out the main parachute, opened as intended. However, due to a failure of an atmospheric pressure sensor, the main parachute failed to deploy. Komarov quickly released the reserve chute, but it became entangled with the drag chute and he crashed into the ground. The unfortunate cosmonaut died instantly. The Soyuz had not really been ready for a test flight, but had been forced into service by the Soviets to keep up the fast-paced space race with America (see Figures 53, 54 and 55).

In 1983, cosmonauts Titov and Strekalov experienced a very fast landing without first having gone into space. They were waiting on top of their rocket to start what should have been Soyuz mission T-10 when a valve in a propellant line failed to close 90 seconds before liftoff. Only one minute before launch, it caused a large fire at the base of the launch vehicle that quickly engulfed the entire rocket. To make matters worse, the automatic crew escape system failed because its electrical wires had been burned through. Luckily, two alert launch controllers in the launch blockhouse manually sent the required commands by radio. Twelve seconds after the fire started, the launch escape rockets ignited, pulling the Soyuz descent capsule from the burning rocket and its enormous load of

FIGURE 53   *Still high above the clouds, a Soyuz capsule gently descends under its parachute.*
[Photo: ESA]

FIGURE 54   *The landing of a Soyuz capsule is softened by a series of small rockets that are ignited just before touchdown, creating a cloud of smoke. Even with these rockets, a Soyuz landing can be quite violent. [Photo: ESA]*

FIGURE 55   *The blackened outer hull of this Soyuz capsule is proof of the intense heat the craft experienced as it reentered the atmosphere. [Photo: ESA]*

highly inflammable propellant. After being subjected to almost 17 g, the crew landed safely some 4 kilometers (2.5 miles) from the launch pad. The rocket exploded seconds after the Soyuz separated. Luckily, the two cosmonauts sustained no injuries from their brief flight of only 5 minutes and 30 seconds.

Coming back from the Moon, the Apollo astronauts had to endure 7 g during reentry, after which they splashed into the sea and had to wait inside their cramped capsule for the recovery team to find them. (Sometimes astronauts had to wait so long that they got seasick owing to the spacecraft rolling on the waves.) After rescuers in scuba gear had opened the capsule door, the astronauts were hoisted into a helicopter to be taken to a Navy vessel. The astronauts of the first few Moon landings were then put inside an airtight camper until it was ascertained that they had not brought back any dangerous lunar viruses or bacteria. Later it was verified that the Moon is as dead as can be, and the quarantine procedure was eliminated.

In comparison, a Shuttle Orbiter's return is much more benign. The astronauts coming back to Earth are only subjected to a maximum of 1.2 g and land softly on a runway, safely on terra firma.

However, returning from orbit is still a hazardous operation; in February 2003 seven astronauts lost their lives when the Shuttle Orbiter Columbia broke apart during reentry. At an altitude of 60 kilometers (37 miles) and a velocity of over 19,000 kilometers per hour (12,000 miles per hour) they died when their vehicle was shattered into pieces and burned up.

SPACEDIVING

An extreme way to return to Earth from orbit – one that would only appeal to the ultimate thrill seekers – is orbital skydiving. Imagine yourself floating out of a spacecraft in a spacesuit, igniting a small rocket to brake and fall back to Earth, deploying an inflatable heat shield, going through a blazing reentry and finishing the last part of the descent under a parachute. A thrilling combination of spacewalking and skydiving.

Part of this has already been done. In 1960, Captain Joseph W. Kittinger was lifted to an altitude of 31 kilometers (almost 20 miles) by a huge balloon containing nearly 3 million cubic feet of helium. To test a new parachute system for escaping from high-altitude aircraft, he jumped from the gondola and deployed a small parachute to stabilize his descent.

Rushing through the rare atmosphere he achieved a maximum speed of 1,149 kilometers per hour (714 miles per hour), slightly less than the speed of sound. Kittinger fell for 4 minutes and 36 seconds before the air got dense enough to allow him to use his main parachute. Captain Kittinger's high-altitude jump record still stands.

In the early 1960s, emergency orbital bail-out systems were developed to enable astronauts to escape from crippled space capsules. One such device, called MOOSE, was developed by General Electric. The MOOSE consisted of a chest-mounted parachute, a folded, 1.8-meter (5.9-foot) diameter elastomeric heat shield and a canister of polyurethane foam. The whole contraption, including the astronaut, weighted no more than 215 kilograms (474 pounds).

FIGURE 56 The spacesuits used in orbit cannot survive a reentry through the atmosphere. The underwear the cosmonauts are wearing in this picture is fitted with cooling lines through which water runs to prevent overheating during heavy work in space. [Photo: ESA]

A spacesuited astronaut would strap the MOOSE to his back and float out of the troubled spacecraft into free space. Pulling the deployment cord would inflate the heat shield with rapid-hardening polyurethane foam, encasing the back of the astronaut in a perfectly form-fitting protective cover. Next, the astronaut would use a small hand-held gas jet device to orient himself in the correct position, then fire a solid rocket motor mounted in the device. After aligning for reentry and putting himself and the MOOSE into a slow spin for stabilization, he would discard the gas gun. Following a ballistic reentry, the astronaut would pull the ripcord of the chest-parachute, which would pull him away from the heat shield. He could then land as a regular parachutist. The astronaut could also decide to leave the shield attached during landing; it could serve as a life raft when landing on water, or absorb the shock of a landing on solid ground. A survival kit with food and water was encased in the polyurethane foam for emergency landings in uninhabited terrain.

Various tests were performed with parts of the system. In one of them, a test pilot jumped 6 meters (20 feet) from a bridge, successfully survived water impact and floated downstream. Other tests confirmed that the foam material could sufficiently protect an astronaut from the intense heat of a reentry. However, the system was never tested in space; the MOOSE project perished together with the cancelled military space stations and the X-20 Dynasoar spaceplane.

New technologies and materials can be used to revive this old idea – not, however, for emergency escapes, but for the ultimate extreme sport thrill (see Figure 56).

PUTTING YOUR FEET BACK ON THE GROUND

How long do you need to get used to 1-g living once you are back on Earth again? In general, the longer you have stayed in microgravity, the longer it will take you. In orbit, blood and other body fluids have pooled in the head and upper torso, only to rush back down toward the feet during atmospheric reentry. This leaves many returning space travelers slightly dizzy.

Because the human orientation system has grown accustomed to floating around in microgravity, astronauts tend to walk in curves shortly after their return. Several Russian cosmonauts who went on long-duration missions, have said that even months after their flight they still occasionally let go of a cup or some other object and are quite surprised when it crashes to the floor. During the first few nights back on Earth, many astronauts feel as if they are constantly rolling over in bed. One astronaut stated that he grabbed the edge of his bed to prevent "falling" onto the floor, while he was actually lying flat on his back all the time. Some astronauts report that while climbing stairs, they imagine the stairs are coming toward their feet, rather than vice versa.

Mark Shuttleworth, the second paying passenger to visit the International Space Station, said that shortly after his flight he was a little unsteady on his feet. He described the feeling as being like a newborn deer adjusting to unfamiliar legs, and said that he kept expecting to be able "to just float" somewhere. The exact reasons for these strange sensations are not yet well understood (see Figure 57).

People who have spend months or even years on board space stations also have to cope with weakened muscles. Even if they trained their body intensively on treadmills and rowing machines during their space mission, it still takes months before they regain their preflight condition.

A less serious, but very obvious effect of spaceflight is that returning long-duration astronauts are a few centimeters longer than when they left. Their spine has expanded owing to the lack of gravity pulling the body down. The effect is only temporary, so buying new clothes is not necessary.

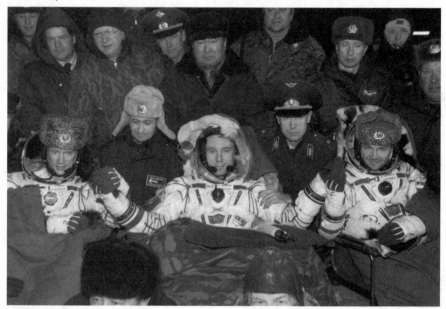

FIGURE 57   *After several weeks or months of microgravity, cosmonauts returning to Earth find it difficult to walk or even stand up. They are therefore put into chairs as soon as they have been assisted out of their capsule. [Photo: ESA]*

SOUVENIRS

What is a holiday without a souvenir? Apart from loads of pictures, space tourists may want to take home some mementos from their flight. Their flight overalls would be obvious things to keep, but they should also be allowed to carry some small personal items with them into space.

Professional astronauts have carried flags, stamps, rings and mission badges into orbit, making them valuable souvenirs and collectors' items. The fact that something has flown in space adds a mystical quality to otherwise quite ordinary things.

When second "Flight Participant" Mark Shuttleworth came back from his ISS trip, he wanted to take the Soyuz capsule he flew in back home to South Africa as a souvenir. Unfortunately for him, the Russians had no desire to sell the spacecraft, but at least he was allowed to keep his spacesuit.

Talking about selling space souvenirs, what about orbital Duty Free sales? Bottles of whiskey would be too heavy to transport into space just to save a couple of dollars on tax money, but small items such as jewelry may be sold on board future space tourism vehicles. Crystals grown in microgravity, where they can become much larger than on Earth, may make for special "Made in Space" jewelry.

Apollo moon rocks are the ultimate twentieth-century space souvenirs. When Apollo moonwalker Harrison Schmitt was asked about this during a recent lecture on his flight, he ensured the public that the astronauts delivered all moon rocks to the scientists only. Considering the amount of effort and money that was required to bring those stones from the Moon, holding any of them back as a personal souvenir could be considered a crime.

However, pieces of Apollo moon rocks have been offered as gifts to heads of states and museums. A small piece that former US President Nixon gave to the Honduras government in 1973 somehow made its way to the market in 1998, but was seized by the US government because they claimed it was stolen. In July 2002 three student employees at NASA's Johnson Space Center were arrested for stealing a 270-kilogram (595-pound) safe full of priceless Apollo lunar samples. They tried to sell their loot via the Internet, but were caught by FBI agents pretending to be potential clients.

There appears to be a market of wealthy collectors looking for pieces of the Moon; the man who tried to sell the Honduras rock was asking $5 million for it. Lunar meteorites, which landed on Earth after been shot into space by meteor impacts on the Moon billions of years ago, are highly

FIGURE 58   Moon rocks, such as this one collected during an Apollo Moon mission, could become popular souvenirs. [Photo: NASA–JSC]

valued too. It is not surprising that many fake moon rocks and lots of worthless dirt, claimed to be Moon dust, have been put on the market.

Once humanity goes back to the Moon and brings more rocks to Earth, the value of the Apollo samples will probably go down. Nevertheless, it will always be special to own something that was brought to Earth by those very first moonwalking pioneers (see Figure 58).

11

TO THE MOON, MARS AND BEYOND

ONCE spaceplane flights have become mainstream holiday trips and chains of space hotels ring the Earth, what will be next? Where will we go for a new vacation thrill? Similar to terrestrial tourism, space tourists will follow in the wake of professional explorers, in this case astronauts, to the Moon, Mars and further into the solar system.

Why? Because even though our scientific satellites and probes can show us views of the planets in a spectrum ranging from Gamma rays to radio waves, the pictures that make newspaper front pages are those made in visible light. We like to see places in space as we would see them with our own eyes if we were there in person. So, why not actually go ourselves?

FLY ME TO THE MOON

What will be the next destination for a space tourist who has "done" low Earth orbit? A logical destination for an expanding space tourism industry is the Moon; it is not very far away, it offers reduced gravity and fascinating landscapes and we already know how to get there (and back).

Other than you would expect after a look through a telescope, the relief on the Moon is rather smooth; the sharp shadows you see are not created by the steep, towering mountains seen in old fantasy drawings, but are caused by the lack of a light-softening atmosphere.

Nevertheless, the Moon is a fascinating place with huge meteor craters, deep ravines and steep hills. Because there is no atmosphere, most of the Moon's surface has been preserved just as it was after the huge meteor bombardments stopped and the craters filled with lava some 3.1 billion years ago. There has never been erosion from wind or water, no plate tectonics to refresh the surface and, apart from a small number of landings, no human influence or pollution. When addressing the appeal of experiencing this non-Earthlike environment, lunar astronaut Alan Bean commented that even though the Grand Canyon is more beautiful than the Moon, it still looks like Earth. In contrast, the Moon presents a completely alien environment that is nothing like our home planet (see Figure 59).

FIGURE 59 *Future space tourists may enjoy hiking trips on the Moon. First aid will need to be adapted to help with the inevitable accidents that will happen. [Photo: NASA]*

Sunrises on the Moon are different from those on Earth: as there is no atmosphere, the Sun appears very suddenly, without the colorful display to which we are accustomed. The thin strip of bright light comes speeding up to you from the horizon, which is only 2.5 kilometers (1.6 miles) away (against 4.7 kilometers or 3 miles on Earth). The golden band is cut by the sharp shadows of distant mountains. You can view as many sunrises and sunsets as you wish by walking back and forth over the sharp border (called the "terminator") between the dark and the light part of the Moon.

Apollo astronauts observed that the color of the lunar surface changes with the sun angle: when the Sun stands high above the horizon, its color is a very bright tanned brown, but when the Sun is low, it looks gray. Even dark soil looks bright with respect to the blackness of space, like light on a dark asphalt road at night.

Other interesting views are caused by Earth and solar eclipses. During an Earth eclipse, when the Moon stands between the Sun and the Earth, you can see the shadow of the Moon moving over the Earth as a dark circular spot. When the Sun is directly behind the Earth, a solar eclipse shows as a bright orange ring around the Earth.

Owing to the lack of wind and rain, the footsteps left in the gray dust by the Apollo astronauts are still as crisp as they were 30 years ago. With space tourism to the Moon, the preservation of the lunar surface would become an important issue because the footsteps of tourists will also remain forever. Can you imagine the first space hooligan destroying Neil Armstrong's first footprint on the Moon? Or someone writing his name in a patch of undisturbed dust near the Apollo 17 landing site, as a kind of everlasting graffiti?

The moonwalkers left lots of equipment on the surface, such as the lower part of their lander, complete moonrovers, scientific instruments and other things such as the medals of two dead cosmonauts and a piece of the Wright Flyer. There are also many uncrewed American and Russian robotic lunar landers to be found. It is not difficult to foresee that these landing sites will one day become both historical and delicate monuments requiring preservation for future generations of lunar tourists. Relatively small but important locations could be covered by unpressurized, transparent domes to keep people and dust out. These domes could be made of glass, manufactured from silicates found in moonrocks and dust.

Apart from historical landing sites, there will also be many unvisited areas on the Moon that we want to leave pristine for scientific or natural heritage reasons. Lunar natural reservations will have to be set up, probably selected by an international organization of scientists. Designated, strictly limited transport paths will have to be set out to prevent footstep and wheel-track pollution (walking trespassers may easily be found by the size and sole-shape of their boots).

The collection of moon rocks by tourists may also have to be limited to preserve the natural environment. An interesting way to do this is currently applied in Hawaii, where a myth has been spread that removing rocks from the volcanoes upsets the volcano goddess and brings bad luck. A similar story about the spell of the lunar goddess Selene (or any of her equivalents from other cultures) may act as a deterrent on the Moon.

Protecting sites of historical and environmental importance on the Moon will probably become the task of some kind of lunar heritage guards. Moreover, just as on space hotels, some type of police body will need to watch over the security of the space tourists, and the formation of a Space Ranger corps may become necessary.

As these rangers will probably need to operate in places that either belong to "all of humankind" or to various nations on Earth, they will need special jurisdiction based on international law. Their job can probably be combined with that of a lunar tour guide, with responsibilities similar to those of today's park rangers in Africa and elsewhere. In this way, they could always be present to protect and take care of the lunar tourists – and to arrest them in case of misbehavior!

Lunar activities

Moving around in the $\frac{1}{6}$-g environment of the Moon is rather easy, once you learn how to hop like a kangaroo and bounce sideways, one leg at a time. You can jump much higher and farther than on Earth, and it is easier than walking. Many activities such as washing, cooking and eating on the Moon would not be very different from normal, and much easier than in microgravity.

Inside pressurized habitats various new lunar sports may be enjoyed, such as flying with wings attached to your arms and low-gravity gymnastics. You can lift weights six times heavier than you can on Earth, although your muscles would get weaker during a long stay on the Moon. You would probably not be able to swim much faster than normal because the density ratio of water versus that of your body stays the same, but professional swimmers would have to optimize their techniques for the higher waves in a lunar swimming pool. Whether distance running records would be broken also remains to be seen; although you can raise your feet six times higher on the Moon, it also takes six times as long before your feet touch the ground again, so, when running, these effects would cancel each other out. Nevertheless, lunar Olympic Games would be spectacular, with athletes lifting 2,800-kilogram (6,200-pound) weights, jumping 9.4 meters (31 feet) high and diving off planks 60 meters (200 feet) high, giving them six times as much time to perform their figures. You jump less than six times higher on the Moon, because your center of mass is already 0.8 meter (2.6 feet) off the ground just before a jump. Leaping 2.4 meters (7.9 feet) up thus actually means you lift yourself 1.6 meters (5.3 feet) high. With the "Fosbury Flop" technique an athlete keeps his or her center of mass 0.2 meter (0.7 foot) under the pole, so to estimate your possibilities on the Moon, you have to subtract 1 meter (3.3 feet) from your jump on Earth, multiply the rest by six, then add 1 meter. Increased "air time" could also make many martial arts even more spectacular.

Outside, on the surface, lunar golf can be played, as Apollo astronaut Alan Shepard did in 1971. The low gravity makes the ball fly six times as far as it would on Earth, and because there is no aerodynamic drag, it would fly even further. The lunar surface also offers an abundance of natural bunkers! However, the necessary moonsuit would probably make your swings more difficult to perfect.

For more extreme forms of sport think of moonbuggy rallies, racing over the rough terrain on balloon tires, flying meters high into the air when driving over a rock and leaving a spectacular trail of dust. No speed limits, no traffic lights, no roads. Several Apollo missions took moon rovers with them, and although the top speed of the battery-driven open car was only some 16 kilometers per hour (10 miles per hour), the low lunar gravity made it look spectacular nonetheless (see Figures 60 and 61).

Lunar skiing may also be a possibility, considering the fine, slippery moon dust and the low gravity.

Flying over the Moon could be another major attraction. Just a short burst of a rocket engine will put a small spacecraft into a parabolic trajectory; like a frog, a rocket vehicle can leap from one point on the surface to another with minimum use of propellant.

For experiencing a holiday location on Earth we not only use our vision, but also the other senses: hearing, smelling and touching make a place more real. You smell the trees, the grass or the ocean, and you feel the rocks under your hands or the sand on the beach running through your fingers. A lunar astronaut is restricted from having such a full experience owing to the lack of an atmosphere and the necessary use of a moonsuit. A partial solution to this incompleteness would be the exceptionally daring plan that science fiction author Larry Niven suggested: be the first to walk on the Moon barefooted! You would be in a pressure suit except for the boots. Tourniquets around the ankles and calves would keep the pressure in the suit. You'd run out of the airlock, leave

FIGURE 60   *The Apollo astronauts used battery-powered moonrovers to drive around. Buggy safaris on the Moon could be major tourist attractions. [Photo: NASA]*

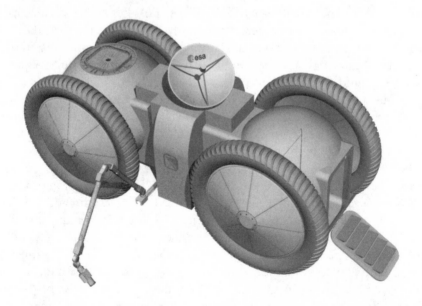

FIGURE 61   *Pressurized rovers such as this could be used to explore the Moon without the need to wear a spacesuit. [Photo: ESA]*

your footprints in the lunar dust and get back inside quickly to have your feet treated for dehydration, frostbite and burst blood vessels. Afterwards a fence would be put around the site, with a sign stating the importance of the footprints. It may be risky, but it's a sure way to make history.

Less dangerous would be the creation of space art on the Moon. When working outside, the use of paint would be impossible because of the extreme temperatures and the lack of air, but alternatives such as special pencils or electronic painting may be used. This would also work in Earth orbit or on Mars, making it finally possible for space artists to work on location.

Speaking about artists, what about souvenirs and shops on the Moon? As these are bound to pop up wherever tourists and hotels are found, it is very likely that we will be able to buy Moon-made artwork, tiny bottles of lunar dust and T-shirts with the slogan "A small step for Armstrong, but a giant leap for me!" However, the price of transportation back to Earth would probably limit the mass and size of souvenirs considerably.

Reaching the moonbase

How would you get to the Moon? Probably you would first get into Earth orbit with a reusable spaceplane, change at a spacestation to a lunar transfer vehicle, which would bring you to another station in orbit around the Moon. There, you would board a lander to take you to a moonbase on the surface. The whole trip could be done in about three days. Going to the Moon using less spacecraft and spacestations, as the Apollo missions did, is certainly possible, but such a system would be less efficient in providing permanent access to the Moon.

On the Moon you would live in habitat modules, covered by thick layers of ground or made of thick walls of lunar concrete for protection against the harsh radiation from the Sun and space.

Solar energy is abundant during the long lunar day – which is half an Earth-month long – and can be turned into storable electricity. Oxygen can be produced from the rocks and ground, while plants can be grown in greenhouses to provide food (experiments on Earth have shown that the lunar dust is an excellent soil for plants). Water is not readily available on the Moon, except perhaps in the deep craters near the poles, where the Sun's heat never reaches the bottom and ice from comets that hit the Moon long ago may still be preserved.

Obviously, such a large and expensive lunar tourism infrastructure of spacecraft and stations is not something we can expect to have before a mature, successful space tourism industry in low Earth orbit is operational.

Cruises to the Moon (without landing) on board a luxury space liner are seriously being studied by Bigelow Aerospace, a new company set up and owned by Robert Bigelow, billionaire and owner/operator of Budget Suites of America, a hotel chain. After his success in the extended-stay accommodations business in Las Vegas, he now turns his visions to the Moon.

The company's objective is a 46,500-square-meter (500,000-cubic-foot) lunar fly-by cruise ship for 100 passengers and a crew of 50. The giant spacecraft concept looks like a dumb-bell with rounded platforms at both ends, designed to hold about 48 staterooms of 3 by 6 meters (10 by 20 feet) with a bath, luxury beds, communications systems and large windows. The complex will rotate once per minute to induce the equivalent of 40 percent of Earth's gravity within the two large modules, while the guests can still enjoy microgravity conditions in the center hub. Partial gravity makes relatively normal showers and toilets possible and will also greatly facilitate dining in the luxurious restaurant, which will serve food of the same quality as ocean cruise ships on Earth.

From special viewing rooms, space tourists will get personal astronomy lessons as the lunar surface flies by underneath. Views from externally mounted telescopes will be shown on screens that cover the walls and ceilings, giving the passengers the feeling of standing on the bridge of the starship *Enterprise*.

In case of emergency, the space cruiser will be able to separate into two self-sustaining parts while there will also be plenty of escape-pods. Long-term exposure of the crew members to high levels of radiation from space is to be solved by using water tanks as shielding.

Passengers will probably be ferried from Earth orbit to the spaceship with rocket shuttles, perhaps with a brief stay at an Earth-orbiting hotel. After boarding, the cruiseship will leave the vicinity of the Earth, fly past the Moon, and return.

Rather optimistically, Bigelow believes that it can be developed for about $2 billion, only a fraction of the price of the International Space Station. To get things off the ground and show he is serious about this, Robert Bigelow is investing $500 million of his own fortune into Bigelow Aerospace, spread over the next 15 years.

A near future possibility for lunar tourism is to land a robot moonrover and have it operated by the general public. With stereo cameras and virtual reality interfaces it would be just like being there. The velocity of the rover cannot be very high, because the video signal would take about 1.5 seconds to reach you, and it would take another 1.5 seconds to relay your steering inputs to the robot.

You could visit Apollo landing sites with a remote-controlled moonrover, or even venture into areas that are too dangerous for astronauts. Science fiction author Larry Niven suggests that the rover could be steered by Jay Leno on television and the project sponsored by, for instance, Toys-R-Us or Sega.

MARS VACATION

Next on the itinerary of a space travel agency would be Mars, a fascinating place with dried river beds, ancient (non-active) volcanoes, polar caps, meteor craters and the largest canyons in the solar system. With 38 percent of the Earth's gravity, it would be easy to move around, even when wearing a spacesuit.

Even though it is further from the Sun than Earth, the red planet receives quite a lot of light and sufficient warmth to sustain a relatively mild climate. As planets go, it is not an extremely warm or cold place; temperatures drop to minus 85 degrees Celsius (minus 120 degrees Fahrenheit) at night, but can rise to just over zero degrees during the day.

Mars has only a thin atmosphere, less than 1 percent of that of the Earth, but it nevertheless has clouds and weather. The wind sometimes picks up the fine, red dust off the surface to create small sand devils or gigantic dust storms that can cover the entire planet.

There is a lot to explore on Mars; although the planet is much smaller than Earth, the amount of land surface is actually larger because it has no oceans (see Figure 62).

Valles Marineris, the planet's vast canyon system, extends over 4,000 kilometers (2,500 miles) and covers one-fifth of the circumference of Mars. It is the largest canyon in the solar system. Some parts run as deep as 7 kilometers (4.3 miles) and as wide as 200 kilometers (124 miles). If Valles Marineris was placed on the surface of Earth, it would stretch from Los Angeles to the Atlantic coast of the USA. The Grand Canyon on Earth is quite small compared to this, with a length of only 446 kilometers (277 miles), a width of 30 kilometers (18 miles) and a depth of 1.6 kilometers (1 mile).

Olympus Mons is the largest mountain in the solar system. It is a shield volcano of 624 kilometers (374 miles) in diameter (approximately the size of Arizona) and 25 kilometers (16 miles) in height, so high that its top is outside the atmosphere. The caldera, the large crater on top of the volcano, is no less than 80 kilometers (50 miles) wide! In comparison, the

FIGURE 62 *Mars is smaller than Earth but does not have any oceans or lakes. The total land surface is slightly larger than that of our planet; there is a lot to explore! [Photo: NASA]*

largest volcano on Earth is Mauna Loa on Hawaii, only 10 kilometers (6.3 miles) high and 120 kilometers (75 miles) across. The volume of Olympus Mons is so enormous that it could contain a hundred Mauna Loas. Science fiction writer Kim Stanley Robinson describes an expedition climbing Olympus Mons in one of his stories: it could be an interesting sport for adventurous mountaineers looking for something less "mundane" than climbing Mount Everest.

However, despite its large polar caps, Mars does not appear to be a good place for winter sports. Observations by NASA's Mars Global Surveyor spacecraft indicate that the snow is rock hard.

No aliens have yet been found on the planet, so a Mars Zoo seems to be out of the question, but a range of uncrewed robotic lander and marsrover missions are planned to look for microbes. The search for fossils from an earlier, warmer and wetter period in Mars' history may become a tourist attraction.

Liquid water must have been abundant on the Martian surface, as shown by the dry riverbeds and the remains of what looks like an enormous ocean once covering a large part of the northern hemisphere. When the planet lost most of its atmosphere because of its low gravity, water could no longer exist in liquid form. However, NASA's Mars Odyssey spacecraft recently verified that there are still large amounts of water ice in the Martian soil, probably enough to create a global Martian ocean, 500 meters (1,640 feet) deep. Water is very important if we ever consider living on Mars: you can drink it, grow plants in it, and make breathable air and propellant (oxygen and hydrogen) for surface vehicles and spacecraft from it. Additional oxygen and methane can be extracted from the atmosphere.

Unfortunately Mars is very far away, about 80 million kilometers (50 million miles) minimum (the actual distance depends on where Earth and Mars are in their orbits around the Sun). With our current propulsion technology, a space cruise ship would take almost 2.6 years for a Mars

FIGURE 63 *Early habitation modules on Mars may look like this. Real Mars bases or even hotels are not likely to be built before the second half of this century.* [Picture ESA]

roundtrip, of which you would spend 1.5 years on the red planet – probably too long for a holiday!

It may take a long time before tourists are able to venture to the red planet, but the first crewed Mars project may be started at any time in the near future. We know how to do it and we have the technology to land people on Mars and bring them back; it's mostly a question of money and political will to commit to such a bold undertaking (see Figure 63).

SPACE COLONIES

Plans for true space colonies, much larger than the space stations described earlier, were first seriously proposed in the early 1970s by Princeton University professor Gerard K. O'Neill. In his book *The Final Frontier: Human Colonies in Space*, concepts of giant rotating wheels and cylinders with diameters of several kilometers are envisioned. They would make mankind a space-based civilization, creating room for Earth's expanding industry and population. O'Neill's space colonies would also offer a chance for our survival in case a large asteroid impact – such as the impact that killed-off the dinosaurs – makes the Earth's environment unlivable for us. (Science fiction writer Larry Niven once remarked, "The dinosaurs died out because they didn't have a space program.")

O'Neill's colony designs are in fact giant greenhouses in which plants and machinery are combined to sustain a healthy environment. Plants would convert the carbon dioxide from breathing people and animals into oxygen, while water, nutrients and other consumables would be recycled. Ideally, the colonies would form closed ecosystems, like mini copies of the Earth itself.

The space colonists would build their space islands from lunar material and space solar power stations (huge satellites that collect the Sun's radiation and beam it down to Earth to be converted into electricity) to make a living and could eventually become independent from Earth (see Figure 64).

Life for the 10,000 or more inhabitants inside these giant spacecraft would be somewhat similar to that on Earth, with people living in a 1-g environment, sleeping in beds and eating regular food in quite normal houses and going to work every day. However, close to the center of the colonies there would be a microgravity environment, useful for large-scale production of special materials. This would also be the ideal place for all kinds of 0-g leisure activities.

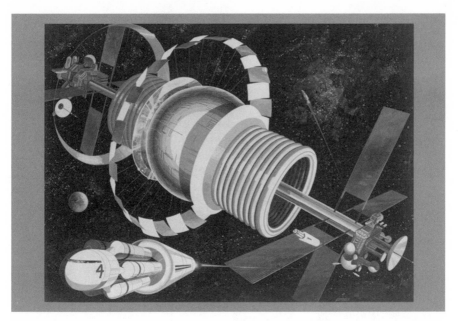

FIGURE 64   *In the far future, people may live and have holidays in giant space colonies such as this. They could be constructed from material launched from the Moon. [Artwork: NASA-ARC/Rick Guidice]*

MORE DISTANT WORLDS

What about the other seven planets? Do they have any suitable locations for future space holiday resorts?

Mercury

Mercury, the small planet closest to the Sun, looks a lot like the Moon. It is gray, dusty and full of craters. It has mountain ranges as high as the Pyrenees on Earth and the gravity is as low as on Mars, but it is doubtful whether those interesting features can compensate for the bad sides of the small planet: intense radiation from the Sun, temperatures ranging from a scorching 415 degrees Celsius (780 degrees Fahrenheit) at the dayside to a freezing minus 170 degrees Celcius (minus 275 degrees Fahrenheit) on the nightside, and no atmosphere. It is also further away than Mars.

Venus

Venus is even worse. Although the planet is named after the goddess of love and was once assumed to be wet, covered with lush vegetation and swamps, the place is actually a hell: atmospheric pressure 90 times higher than on Earth, a surface temperature of 470 degrees Celcius (880 degrees Fahrenheit), an atmosphere consisting mostly of carbon dioxide and clouds of sulfuric acid from which a nasty acid rain falls down. It is a planet suffering from a runaway glasshouse effect. Club Med would not be interested.

The giant planets and Pluto

Beyond Mars is the realm of the giant gas planets Jupiter, Saturn, Uranus and Neptune.

Jupiter is the largest planet in the solar system, with a maximum diameter of 142,800 kilometers (88,700 miles), 22 times that of the Earth. It is a beautiful planet, colored with brown, yellow, orange and red cloud bands circling the planet parallel to its equator. Unfortunately, it does not have a solid surface to walk on because Jupiter consists solely of gas. It also emits deadly radiation.

It has, however, a series of fascinating moons, the largest of which are Ganymede, Callisto, Europa and Io. Ganymede is larger than our own Moon, and even larger than the planet Mercury. It has mountains the size of those on Earth. The moon Callisto has a surface of dark, muddy ice filled with craters that has not changed since the birth of the solar system. Icy Europa is especially interesting. It has an icecap that is not only some kilometers thick but envelops the whole planet, probably covering a giant ocean that may harbor life. The pizza-colored moon Io would also be an essential stop on a cruise through the Jupiter system. Tidal forces, caused by the giant planet which it orbits, knead Io and so heat up its interior. This results in giant volcanoes that spew out sulfuric gasses into space. It is a pity that Jupiter is so far away – about 630 million kilometers (390 million miles) minimum, when the planet is at its closest point to Earth.

The planet Saturn is best known for its magnificent ring system, made up of millions of individual bands of rocks and dust. Its largest moon, Titan, has a nitrogen atmosphere and may be covered by methane lakes, snow and ice. It may even rain methane on Titan. We will not know for sure what the surface actually looks like until the Huygens probe of the European Space Agency lands on the moon in January 2005. It is currently underway, attached to NASA's Cassini spacecraft that will further explore Saturn and its moons (see Figure 65).

FIGURE 65   In the far future, space tourists may be able to visit Saturn and marvel at its beautiful rings. [Photo: ESA]

The planets Uranus and Neptune have less interesting surfaces than Jupiter and Saturn and are very far away. Even their beautiful moons will not convince us to go there before a revolution in propulsion technology has taken place to drastically reduce interplanetary travel time.

Pluto is the most distant planet. It is small, extremely cold and very dark. Astronomers actually argue whether to call it a planet or an oversized asteroid. No spacecraft has visited little Pluto yet, but from what we know it does not seem to be a promising vacation spot.

Asteroids and comets

Apart from the nine main planets, the solar system is full of asteroids and comets. Asteroids are metallic, rocky, often potato-shaped bodies without atmospheres. They orbit the Sun but are too small to be classified as planets. Tens of thousands of them congregate in the so-called main asteroid belt, which is a vast, doughnut-shaped ring located between the orbits of Mars and Jupiter. The largest asteroid is Ceres, about 1,000 kilometers (600 miles) in diameter, but the smallest are the size of pebbles. Only 16 asteroids have diameters of 240 kilometers (150 miles) or more.

Comets are large bodies of water ice, dust and rocks. They have highly elliptical orbits that repeatedly bring them very close to the Sun, where they develop beautiful tails of gas as they partly evaporate in the heat.

Both asteroids and comets could be interesting to visit, as their extremely low gravity may have created all kinds of fascinating landscapes.

Walking on them is, however, very tricky: a small jump will be enough to launch yourself into space!

The constraints of distance

A major problem with all these destinations is their distance. If we shrink the Sun to the size of an orange, Earth is the size of a grain of sand at a distance of 10.5 meters (34 feet), Mars is 16 meters (52 feet) away; Jupiter is the size of a small marble at 54.5 meters (178 feet) and Pluto is an invisible speck of dust 414 meters (1,359 feet) from the Sun. In this model, spacecraft move at only a few tens of meters per year. Large spacecraft with similar onboard facilities as cruiseships may be the only way to utilize the vast tourism potentials of the solar system.

What if we go beyond the solar system's boundaries and enter the vastness of interstellar space? The London Planetarium has a show where you become an intergalactic tourist on a cruise through the universe. After settling back in the reclined chairs, the lights go out and a large projection of the starry sky is shown on the cupola above. A launch from Mars brings you into space, where you can have a look at other galaxies, massive exploding stars called supernovas and venture near a black hole where the gravity is so high that even light cannot escape. The imaginary spaceship that the planetarium becomes uses wormholes to get quickly from one point of the universe to another.

Maybe such a trip will be possible in the far, far future, but not before we learn to travel faster than light: the closest star is 4.2 light-years away. This means that traveling at the speed of light, it would still take you 4.2 years to get there. Any other interesting interstellar destination is much further away. This is part of the appeal of space travel: wherever we go or whatever we achieve, we will always seem to be at the beginning, with much, much more to explore ahead of us.

AT THE SPEED OF LIGHT

For reaching destinations at the edge of our solar system and beyond, space tourists will need means of fast transportation that are now still highly hypothetical.

One of the remote possibilities on every space cadet's wish list is the use of antimatter. As far as we know, antimatter does not exist in nature, but it can be produced in large particle accelerators such as those of CERN in

Switzerland and Fermilab in the USA. The quantities that can be obtained are, however, not staggering, only a few antimatter hydrogen molecules in total.

Once produced, it is very difficult to store antimatter particles because they annihilate when they come in contact with normal matter. This is also their benefit: when antimatter and normal matter particles collide, both are completely transformed into energy. The resulting heat can be used to create hot gasses to expel through a rocket nozzle.

Only a small amount of antimatter, kept inside an electromagnetic containment field, is sufficient to create an enormous amount of energy: 1 kilogram of antimatter and normal matter provides 10 billion times more energy than the combustion of 1 kilogram of hydrogen and oxygen. As small an amount as 71 milligrams of matter and antimatter contains the same amount of energy as all the propellant in the Space Shuttle External Tank. In an antimatter launch vehicle, the antimatter tanks could be minuscule, but large amounts of normal propellant would still be required to provide the mass to be expelled through a rocket nozzle.

An even more effective, but also completely hypothetical means of getting into space is antigravity. It is often used in science fiction but no one knows if it is possible. Modern physics theories assume that gravity forces are caused by interference patterns of so-called gravity waves. If we could generate gravity waves in a controlled manner, it might be possible to cancel out or reverse gravity forces. It is somewhat like antisound technology, where noise is suppressed by creating superimposing sound waves that are mirror images of the noise sound waves.

Entering the realm of popular science fiction such as the "Star Trek" TV series, the possibility of teleportation appears to be the ultimate means of transportation: your body is scanned while being broken down to its basic atoms, after which the information is sent to another place where the process is performed in reverse order. In this way, a replica is produced at the receiving location.

Teleportation is an idea that was first used in the original "Star Trek" series as a means of avoiding costly spacecraft mockups and special effects shots of landing spacecraft, but recently it seems there may actually be some science in this fiction.

In 1993, a group of scientists came up with a theory on quantum teleportation, according to which it would indeed be possible to teleport an object, but only by destroying the original in the process. The destruction of the original object is an unavoidable result of the scanning process at quantum level. Recently scientists have actually been able to teleport the quantum state of a single photon, in effect teleporting the particle and proving the correctness of the quantum teleportation theory presented in 1993.

But teleporting a person is different from teleporting a single atomic particle. First of all, you would probably not need to scan people down to quantum level; just knowing the arrangement and position of each molecule would be sufficient. This would also imply that destruction of the subject is not necessary, so you would end up with a replica of the person at another place rather than a teleportation if you didn't do anything about it.

In the "Star Trek" series, almost every way teleportation could go wrong has been explored; from the accidental merging of two travelers to the messy effects of mixed-up teleportation signals causing random relocations of body parts.

Another problem addressed in one of the TV episodes is the inadvertent creation of an exact copy of a person while also the original remains in existence: there is one at the point of origin and another at the destination. In an instant, you would have a twin sister or brother! This would be possible because the teleported replica is not necessarily created from the actual material of the original, but perhaps from atoms of the same kind arranged in exactly the same pattern.

What could make the whole thing impossible is the fact that the human body is very, very complex. The information on "which atoms are where" would amount to hundreds of billions times hundreds of billions of gigabytes. Even with the best optical fibers it would take more than a hundred million centuries to transmit all that information from one place to another.

So, whether teleportation of people instead of a single photon will ever be possible remains to be seen.

In the "Star Trek" series teleportation can only be used over limited distances; for long travels the crews of Star Fleet still depend on spacecraft (which were also more suitable for flashy dogfights and spectacular explosions). These work by the highly hypothetical means of warping space-time. If you imagine three-dimensional space as a two-dimensional sheet of paper with two dots on it, you can either go from one dot to the other the long way, or fold (warp) the paper so the dots are near each other and the distance is dramatically reduced. That is basically the idea behind the "Warp Drive" used by the fictitious starships. As strange as it may sound, the warping of space-time is actually an established, proven concept of cosmology. In fact, our whole universe is warped, although it is not yet clear how much.

Maybe the most outlandish idea for traveling through space is the use of wormholes. Famous physicists Stephen Hawking and Kip S. Thorne speculated that a black hole (a collapsed star with a gravity field so concentrated that even light cannot escape) could connect one area of the universe with another through a kind of tunnel called a wormhole. They could theoretically be used to travel through both space and time, and maybe even connect our universe with a completely different one. Because of this they are, not surprisingly, the subject of many works of science fiction. The great promoter of astronomy, Dr Carl Sagan, wrote about an alien race who created a kind of subway system of wormholes in his book *Contact*.

12

THE ROAD
AHEAD

IN this book we have seen that space tourism is no longer a fantasy that only exists in science-fiction novels: it is becoming reality. Two pioneering tourists have already flown into space. In contrast to traditional astronauts, they paid for their flight from their own pockets and went primarily for the adventure, to experience microgravity and the beautiful sight of our Earth from orbit.

Several companies are now offering flights into space on board Russian Soyuz spacecraft, although ticket prices still limit the market to a small group of adventurous multi-millionaires. However, these same companies are also offering "nearly-space" tourism experiences such as microgravity parabolic flights, astronaut training and high-altitude jet fighter flights to a larger public.

Flying to the edge of the atmosphere and experiencing minutes of weightlessness may soon become possible through the suborbital vehicles competing for the X-Prize. While ticket prizes will still be steep, they are expected to be in the hundred thousands of dollars rather than in the tens of millions needed for a full orbital flight.

These developments are similar to what is happening in other areas of adventure tourism. Tourists now take deep-sea trips in submarines that, until recently, could only be used by scientists or visit the North Pole on board nuclear icebreakers that were originally designed for professional expeditions only. Cold war jet fighters are now for sale as private planes, and parachuting has developed from a risky military delivery and escape system into a grown-up sport.

Because of these developments, more and more people are getting used to the idea that privatized spaceflight is no longer impossible.

We now have more than 40 years of experience in human spaceflight. We know that people can eat, sleep, wash, work and relax in space. We can fairly accurately predict how someone's body is going to behave in microgravity, and at least for short flights we know how to sufficiently limit muscle deterioration and bone loss.

Although professional astronauts need to have special skills and be of above-average health, we have discovered that flying in space is not the exhausting, psychologically demanding experience we once thought it was. From a physical point of view, virtually anyone can make a space flight.

We currently have the technology to launch people, keep them comfortable in orbit, and bring them safely back home. We have launched hundreds of crewed spacecraft. We can also build space stations that enable people to live for more than a year in orbit. While spaceflight is still not exactly routine, Space Shuttles and especially Soyuz spacecraft are leaving the Earth on a regular basis and usually fulfill their missions without major problems. Having become familiar with the hazards of spaceflight, we are getting ever better in handling and mitigating dangers.

Early space tourists and professional astronauts have taught us that flying in space is an extraordinary experience that cannot be compared to anything on Earth. Weightlessness, the view and the excitement make a trip into orbit not just an unforgettable event, but have the ability to truly change one's view of the world.

As a result, many "ordinary" people are now wishing to make a spaceflight of their own, and are willing to pay serious money to see their dream come true. The potential space tourism market is huge, if only we could open it.

As far as excitement and entertainment are concerned, the sky is definitely not the limit for space tourism. Large space hotels, microgravity sport facilities and trips to the moon and further into the solar system are possibilities for endless expansion.

COST AND SAFETY

Two major problems stand in our way to large-scale, economically viable space tourism: money and safety.

Spaceflight is still very expensive. This is partly because of its challenging nature: the extreme temperatures, velocities and pressures involved push our technological capabilities like no other type of transportation. Sophisticated, high-quality and therefore expensive equipment is required. An enormous amount of preparation is needed before a launch vehicle can lift off, which is a very different routine from the use of ordinary airplanes.

Another reason for the high cost is that human spaceflight has always been exclusively based on government support and on hardware and procedures derived from military programs. These organizations do not tend to make anything efficient and inexpensive.

The second issue is safety. Although our human spaceflight record is impressive, the number of people and missions flown is still relatively low and the risks are high. Our spacecraft can still be considered immature, and are far from perfected prototypes when compared to commercial and military airplanes.

When Lindbergh crossed the Atlantic, he was exposed to considerable danger in an experimental airplane that could not be guaranteed to stand up to the task. By now, intercontinental flights are safe and routine and hundreds of planes are crossing the oceans every day. In contrast, a crewed spaceflight is still something extraordinary that reaches the papers; astronauts and cosmonauts put their lives at risk each time they go up. For mass space tourism, spacecraft need to become more like aircraft in reliability, maintainability and safety, but this is no easy task.

So, what is going to make space launches affordable and safe, and thereby mass space tourism a reality? Until recently, most people thought that government space agencies would be our best hope. NASA, ESA (Europe) and JAXA (Japan) and their industrial contractors know how to develop launchers and crewed spacecraft, and have the money to develop the necessary technology. However, their primary goals are scientific exploration and technological development, not the setting up of commercial businesses.

Their commercialization of the International Space Station, by offering onboard experiment facilities for rent to private companies, is not really getting off the ground. The environment in which the space agencies have to work, with its entanglement of political, industrial and scientific demands and constraints, complicates space commercialization. This is even more true for space tourism, which also suffers because it is not being taken very seriously by these space agencies.

Furthermore, America's and Europe's human spaceflight plans are currently aimed at the Moon and Mars. The development of spacecraft and space stations for tourism is not on their priority list.

An exception is the Russian Space Agency, Rosaviakosmos, which has embraced space tourism as a quick means of getting desperately needed additional funding. Anything seems to be negotiable and their rules are flexible. The fact that Russian salaries, and therefore launch prices, are relatively low also helps, and it is not surprising that western companies like MirCorp are working with Rosaviakosmos rather than NASA, ESA or JAXA.

In an optimistic scenario, Rosaviakosmos may decide that space tourism is their main space market of the future and the only certain way of supplementing its dwindling budgets. They may decide to further exploit their current leading position in the space tourism business, and focus their future developments on this new market.

Historically, means of transportation such as ships, trains and planes have been kick-started with government support, but made economical by private industries. We may need a similar evolution if space tourism is to become a profitable business.

This process has now started, as private companies are working on advanced, yet simple, and low-cost rocket aircraft to attack the space tourism barrier from the ground up. Importantly, they do this without any direct government support or space agency involvement. Their suborbital spaceplanes may start a sufficiently large space tourism market to enable the development of more advanced systems and, ultimately, fully commercial orbital spacecraft.

This may be accelerated by extremely rich investors, such as those helping some of the X-Prize competitors. These people may put some of their billions into the development of orbital space tourism vehicles without demanding huge profit in the short term.

However, there are also risks to the early development of space tourism that threaten it from growing up.

Reusable launcher technology development is going very slowly. Owing to the low demand for launches, the space agencies do not consider the development of efficient reusable spaceplanes a priority. For its human Moon and Mars exploration plans NASA is seriously considering the use of Apollo-type systems: huge, expendable launchers and non-reusable capsules. This refocus could deprive space tourism businesses of the efficient, reusable spaceplane technology they need.

Without this to start with, it will be difficult to make privatized flights into orbit cheap and safe enough to open a sufficiently large market. If space tourism ticket prices remain in the order of tens of millions of dollars, the small market of a few hundred adventurous multi-millionaires will soon be depleted.

The other way of working toward orbital space tourism, with X-Prize competitors making ever larger and more powerful spaceplanes, could end prematurely by a couple of serious failures. Disasters taking the lives of X-Plane crews, especially if there are paying tourists among them, could scare away potential customers and investors. Even if such things do not happen, the operation of X-Prize vehicles may turn out to be too difficult and too expensive. In addition the scaling up to fully orbital spacecraft may technologically be too steep a hurdle to take.

However, such adversities may stall the development of space tourism for some time, but it is not very likely that it will stop it forever. The first space tourists have already flown and more are following; the doors to space have opened and it will be hard to close them again. People have been enticed to fly ever higher and to experience what was until now only available to a select club of professional astronauts. This strong market push may ultimately be the deciding factor in making mass space tourism a reality.

BE A SPACE TOURIST NOW

A song written for a James Bond movie by the band Garbage says: "The world is not enough, but it is such a perfect place to start." Even though real space trips and suborbital launches may not be within the reach of most people in the near future, there are many things you can do now, right here on Earth, that are related to space tourism.

Cosmonaut training holidays and parabolic and high-altitude jetfighter flights have already been discussed, but even these are quite expensive. Instead, many more affordable space-related tourism activities are available, such as space camps, trips to launch sites, visits to space museums and building and launching model rockets. Visit an astronomical observatory or planetarium, or buy a telescope; staring at the night sky is fascinating. Looking at the Moon, even with simple binoculars, is an impressive experience and with some imagination it feels as if you are in orbit around it.

IMAX movies, shot in space and projected on a huge screen, give a taste of the sensations an astronaut enjoys. For more physical experiences you could try scuba diving or parachuting. Floating under water, having neutral buoyancy, breathing oxygen from a tank and watching a fascinating, almost alien world through a mask has many similarities to spacewalking. Parachuting gives you the sensation of speed and freedom, and during free-fall you are actually weightless.

Furthermore, don't forget that we are actually all space travelers, dashing at 30 kilometers (18 miles) per second around the Sun on board a big blue spacecraft, while the whole solar system circles the center of the Milky Way at an incredible 217 kilometers (135 miles) per second. Our ship carries everything we need to survive: water, a breathable atmosphere, climate control, radiation protection, living space, and a huge crew. Spaceship Earth is an amazing place. Explore it!

BIBLIOGRAPHY

Books and reports

Aldrin, B. and Barnes, J. *The Return*, Tor, Tom Doherty Associates, LLC, New York, USA, 2000.

Ashford, D. and Collins, P. *Your Spaceflight Manual: How you Could be a Tourist in Space within Twenty Years,* Eddison Sadd Editions Limited, London, UK, 1990.

Ashford, D. *Spaceflight Revolution*, Imperial College Press, London, UK, 2002.

Berinstein, P. *Making Space Happen: Private Space Ventures and the Visionaries Behind Them*, Plexus Publishing, Medford, USA, 2002.

Breuer, W.B. *Race to the Moon: America's Duel with the Soviets*, Praeger Publishers, Westport, USA, 1993.

Burrough, B. *Dragonfly: NASA and the Crisis Aboard Mir*, HarperCollins Publishers, New York, USA, 1998.

Couper, H. and Henbest, N. *The Planets*, Pan Books Ltd, London, UK, 1985.

Doran, J. and Bizony, P. *Starman, The Truth behind the Legend*, Bloomsbury Publishing, UK, 1998.

Dyson, G. *Project Orion: The True Story of the Atomic Spaceship*, Henry Holt & Company, Inc., USA, 2002.

Frimout, D. and Hendrikx, S. *Op zoek naar de blauwe planeet*, CODA, Belgium, 1993.

Godwin, R. *X-15, The NASA Mission Reports*, Apogee Books, Burlington, Canada, 2000.

Godwin, R. *Rocket and Space Corporation Energia, The Legacy of S.P. Korolev*, Apogee Books, Burlington, Canada, 2001.

Godwin, R. *Dyna-Soar, Hypersonic Strategic Weapons System*, Apogee Books, Burlington, Canada, 2003.

Hartmann, W.K., Sokolov, A., Miller, R. and Myagkov, V. *In the Stream of Stars: The Soviet/American Space Art Book*, Workman Publishing, New York, USA, 1990.

Houston, A. and Rycroft, M. *Keys to Space: An Interdisciplinary Approach to Space Studies*, International Space University, McGraw-Hill, USA, 1999.

ISU SSP2000 Space Tourism Design Project Team. *From Dream to Reality*, International Space University, Valparaiso, Chile, 2000.

Koppeschaar, C. *De Maan*, Uitgeverij J.H. Gottmer/H.J.W. Becht BV and Carl Koppeschaar, The Netherlands, 2001; English (older) version: *Moon Handbook: A 21st Century Travel Guide*, Avalon Travel Publishing, USA, 1995.

Larson, W.J., Pranke, L.K., Connoly, J., Giffen, R. *et al. Human Spaceflight Mission Analysis and Design*, McGraw-Hill, USA, 1999.

McElyea, T. *A Vision of Future Space Transportation: A Visual Guide to Future Spacecraft Concepts*, Apogee Books, Burlington, Canada, 2003.

Meier, T.A. *Assessment of Unconventional Space Transportation System Candidates, ESTEC WP 1966*, ESA-ESTEC/IMT-TSA, Noordwijk, The Netherlands, 1997.

Nieuwenhuis, H. *Artis in de Ruimte*, Uitgeverij Van Wijnen, Franeker, The Netherlands, 1988.

O'Neill, G.K. *The High Frontier: Human Colonies in Space*, Uitgeverij J.H. Gottmer, Haarlem, The Netherlands, 1978.

Owen, D. *Into Outer Space*, Quintet Publishing Limited, London, UK, 2000.

Reichhardt, A. (editor) *Space Shuttle: The First 20 Years*, DK Publishing Inc., New York, USA, 2002.

Smolders, P. *Wonen in de Ruimte, Handboek voor Ruimtereizigers*, Unieboek, Weesp, The Netherlands, 1985.

Smolders, P. *De zwaartekracht voorbij, Veertig Jaar Ruimtevaart*, Schuyt & Co., Haarlem, The Netherlands, 1997.

Steele, A. *Clark County, Space*, Ace Books, New York, USA, 1990.

Steele, A. *Sex and Violence in Zero-G*, Meisha Merlin Publishing, Decatur, USA, 1999.

Stine, G.H. *Halfway to Anywhere*, M. Evans & Company, New York, USA, 1996.

Stine, G.H. *Living in Space: A Handbook for Work and Exploration Stations Beyond the Earth's Atmosphere*, M. Evans & Company, New York, USA, 1997.

Tuinder, R. *Ruimtelijk Denken*, Guide to "International Documentary Festival Amsterdam goes Space", Cinema Digitaal, Amsterdam, The Netherlands, 2002.

Wendt, G. and Still, R. *The Unbroken Chain*, Apogee Books, Burlington, Canada, 2001.

White, F. *The Overview Effect: Space Exploration and Human Evolution*, Houghton Mifflin Company, Boston, USA, 1987.

Yeager, C. and Janos, L. *Yeager*, Bantam Books, USA, 1985.

Papers and articles

Abitzsch, S. "Global market scenario of a space tourist enterprise", International Symposium on Space Tourism, March 20–22, Bremen, Germany, 1997.

Bremner, C. "Three-seater powers up for cheap space trip", *The Times*, UK, May 8, 2003.

Burgess, C. and French, F. "Only males need apply: The Lovelace Women, and the Not-So-Right Stuff", *Spaceflight*, vol. 41, no. 1, UK, 1999.

Carreau, M. "New heights for partners taking adventure travel", *Houston Chronicle*, USA, March 15, 2002.

Chang, K. "Not science fiction: An elevator to space", *New York Times*, USA, September 23, 2003.

Chang, K. "Thinking beyond the shuttle", *New York Times*, USA, October 2, 2003.

Collins, P. *et al.* "Demand for space tourism in America and Japan, and its implications for future space activities", *Proceedings of the 6th ISCOPS, AAS*, vol. 91, 1995.

Collins, P., Fukuoka, T. and Nishimura, T. "Orbital sports stadium", *Space 2000*, Albuquerque, USA, 2000.

Davis, J.R., Jennings, R.T. and Beck, B.G. "Comparison of treatment strategies for space motion sickness", IAF/IAA-91-554, 42nd Congress of the International Astronautical Federation, Montreal, Canada, 1991.

Fawkes, S. "Space tourism, and end of the century review", *Spaceflight*, vol. 42, no. 2, UK, 2000.

French, F. "Citizen explorer: The Byzantine odyssey of Dennis Tito", *Spaceflight*, vol. 44, no. 4, UK, 2002.

French, F. "First British ride into space", *Spaceflight*, vol. 44, no. 5, UK, 2002.

Gregory, W. "Barfology. What scientists haven't solved and hot-shot pilots won't talk about", *Air and Space*, vol. 17, no. 1, USA, 2002.

Hodierne, R. "Space oddity", *Washington Post*, USA, December 9, 2001.

Koelle, D.E. "Technical assessment of the minimum 'cost per flight' potential for space tourism", International Symposium on Space Tourism, March 20–22, Bremen, Germany, 1997.

Landis, G. "Pathfinder – a personal retrospective", *Spaceflight*, vol. 44, no. 8, UK, 2002.

Lyndon B. Johnson Space Center. *NASA Facts: Living in the Space Shuttle*, FS-JSC-95(08)-001, NASA Lyndon B. Johnson Space Center, Houston, USA, 1995.

Morris, W.D., White, N.H. and Ebeling, C.E. "Analysis of Shuttle Orbiter reliability and maintainability data for conceptual studies", AIAA 96-4245, AIAA Space Programs and Technologies Conference, September 24–26, Huntsville, Alabama, USA, 1996.

Owens, L. "Space tourism prospects increasing", *Spaceflight*, vol. 45, no. 1, UK, 2003.

Pelt, M.O. van. "Orbital space tourism: Affordable ticket prices in the near future?", BIS Symposium 2001 – The Popular Commercialisation of Space, London, UK, September 19, 2001.

Persson, M. "Jongensdroom gaat zuurstof happen", *De Volkskrant*, The Netherlands, April 3, 2004.

Rowe, W.J. "The reservoir", *Spaceflight*, vol. 45, no. 2, UK, 2003.

Sample, I. "Earth Eleven: In space no one can hear you scream at the referee", *New Scientist*, July 2000.

Scott-Scott, J., Harrison, M. and Woodrow, A.D. "Considerations for passenger transport by advanced spaceplanes", IAA-99-IAA.1.3.06, 50th International Astronautical Congress, October 4–8, Amsterdam, The Netherlands, 1999.

Stone, B. "Let's go to space!", *Newsweek*, October 13, 2003.

Traa, M. "Stoomcursus astronaut", *Kijk*, no. 5, The Netherlands, May 2002.

Zheleznyakov, A. "Rumours, innuendo and gossip: Discovering the truth behind Russia's early space programme", *Spaceflight*, vol. 44, no. 11, UK, 2002.

Zubrin, R.M., Baker, D.A. and Gwynne, O. "Mars Direct: A simple, robust and cost effective architecture for the space exploration initiative", AIAA-91-0328, 29th AIAA Aerospace Sciences Meeting, Reno, Nevada, USA, 1991.

Webpages

Braunstein, S.L. *A Fun Talk On Teleportation.*
www.research.ibm.com/quantuminfo/teleportation/braunstein.html, 1995.

Collins, P. *Space Future.*
www.spacefuture.com, 2002.

Delio, M. *Sex That's Out of This World.*
www.wired.com/news/technology/0,1282,39977,00.html, 2000.

European Space Agency main website:
www.ESA.int, 2002.

Incredible Adventures website:
www.incredible-adventures.com, 2002.

Kridler, C. *Trio Recalls Intense Return.*
www.floridatoday.com/columbia/columbiastory2A51792A.htm, 2004.

Lefcowitz, E. *Retrofuture: Let Them Eat Fake!*
www.retrofuture.com/spacefood.html, 2001.

Lunar Architecture website:
www.lunararchitecture.com, 2003.

MirCorp website:
www.mir-corp.com, 2002.

Niven, L. *Making Somebody Pay.*
Special to SPACE.com website, 14 September 2000.

NASA main website:
www.nasa.gov, 2002.

Scaled Composites website:
www.scaled.com, 2003.

Space Adventures website:
www.spacevoyages.com, 2002.

Space Island Group website:
www2.dowco.com/spaceisland/index.html, 2002.

Starchaser Industries website:
www.starchaser.co.uk

Tryst, O. *NASA shuttle astronauts conducted sex experiments in space* [Internet hoax].
www.snopes2.com/sex/tattled/shuttle.htm, 2000.
Wade, M. *Encyclopedia Astronautica.*
www.astronautix.com, 2001.
Wakefield, J. *Book Me a Double – With a View of Venus.*
www.wired.com/wired/archive/8.01/rocketman.html, 2002.

Video

Frederickson, L. and Bowman, R. (producers). *Inside the Space Shuttle,* Discovery Channel video, USA, 1998.

INDEX

A citation in **boldface** *is a reference to an illustration.*

213